M

PATIENT NUMBER ONE

PATIENT NUMBER ONE

▷ RICK MURDOCK AND DAVID FISHER

A TRUE STORY OF HOW ONE CEO

TOOK ON CANCER AND BIG BUSINESS

IN THE FIGHT OF HIS LIFE

CROWN PUBLISHERS   NEW YORK

Copyright © 2000 by David Fisher and Rick Murdock

All rights reserved. No part of this book may be reproduced or transmitted in any form or by any means, electronic or mechanical, including photocopying, recording, or by any information storage and retrieval system, without permission in writing from the publisher.

Published by Crown Publishers, 201 East 50th Street, New York, New York 10022. Member of the Crown Publishing Group.

Random House, Inc. New York, Toronto, London, Sydney, Auckland
www.randomhouse.com

CROWN is a trademark and the Crown colophon is a registered trademark of Random House, Inc.

Printed in the United States of America

Design by Barbara Sturman

Library of Congress Cataloging-in-Publication Data

Murdock, Rick.
    Patient number one : a true story of how one CEO took on cancer and big business in the fight of his life / by Rick Murdock and David Fisher.—1st ed.
        p.   cm.
    1. Murdock, Rick—Health. 2. Lymphomas—Patients—Washington—Seattle—Biography. 3. Cell separation. 4. Biotechnology industries—Washington—Seattle.
I. Fisher, David, 1946– II. Title.
RC280.L9 M87 2000
362.1'9699446'0092—dc21
    [B]   99-086763

ISBN 0-609-60391-4

10 9 8 7 6 5 4 3 2 1

First Edition

To my family, without whom this story could not have been told:
To my wife, Patricia, for her unwavering support, dedication,
and determination. She gave me the strength to keep moving
forward every day.
To my sons, Jamie and Benjamin. By refusing to accept defeat,
they made it easy to fight on.

To the people of CellPro, who accomplished so much. I am
forever grateful for their hard work, dedication, and loyalty.

To the men and women who live and work on the front lines of
the intersection between Science, Technology, and Medicine.
They provide the daily miracles that push back the frontiers of
disease treatment and patient care.

— RICK MURDOCK

With all my love to my wife, Laura.
How grateful I am to have met her.

— DAVID FISHER

ACKNOWLEDGMENTS

In telling this story we received the assistance of many people. We remain extremely grateful to each of them:

For graciously giving of their time and memories, we sincerely would like to thank Nicole Provost and Frank Fay, Sharon Adams, Stan Corpuz, Dr. Joseph Tarnowski, Dr. Oliver Press, Sherri Bush, Josh Green, Joe Lacob, Dr. Richard Miller, Larry Culver, Dr. Cindy Jacobs, Dr. Amy Sing, Professor Martin Adelman and Mark Handfelt.

For editorial assistance we would like to thank Joanne Curtis.

We deeply appreciate the contribution of reporter Bill Richards, whose hard work first brought this story to the attention of the American public.

This book simply would not have been possible without the efforts of our editor, Kristin Kiser. It is very difficult to accurately convey her impact on this book, but her vision is there on every page.

Finally, we cannot emphasize enough how grateful we are to Crown Publishers. In a climate in which publishers are too often timid, the management of Crown showed great courage. The publisher encouraged us at every opportunity to tell this story the way it needed to be told and the company deserves much credit for that.

CONTENTS

INTRODUCTION

We were sailing on the edge of the wind, nine hours out of Nantucket en route to safe harbor in Newport. It was a day from a sailor's song; the sky was high and blue, the wind was steady, and if you looked far enough you could see beyond the horizon almost into tomorrow. My wife, Patricia, was at the helm, steering a course across Buzzards Bay. My oldest son, six-year-old Jamie, was down below in the care of my mother. I was in the cockpit with Patty and our four-year-old, Ben, trimming the sails and checking the charts.

We had caught the wind early that morning and never let go, sailing through the day. As the late-afternoon sun dropped, it cast a shimmering golden line on the surface. Our boat, the *Mon Pays Bleu,* a thirty-seven-foot Hunter, under full sail, was slicing sweetly through gentle swells.

There are on occasion perfect days at sea, days when the wind blows true and everything aboard is shipshape. When the world works. This was one of those days. As I began to trim the mainsail, I paused to savor the moment. The wind was at our back, we were riding the swells, a single shackle was clanging rhythmically against the mast. For a brief time I was lost in the sailor's reverie. I was filled with joy. I was with the people I loved most, on a boat powered by the wind and under my complete control. We were racing the sunset through deep water, in harmony with nature.

I relaxed. For an instant I had lowered my guard, so when I glanced casually at the depth sounder, I was stunned at what I saw: the bottom was coming up fast to meet us. I had mistaken beauty for security, the mistake that has doomed mariners since man first set sail. I had been lulled into believing I had control of the day. Patty was at the helm.

I took a deep breath, turned to her, and said, "It's getting pretty shallow. We must be getting too close to the reef. Let's tack ou—"

At sea a lifetime can pass in an instant, or an instant can seem to last a lifetime. Nine tons of boat plowed into the submerged reef at about five knots. The screech of our lead keel driving into the rock was almost deafening. Within a few feet we stopped dead in the water. Around me, everything seemed to be happening in slow motion. I saw Ben go flying down the companionway. When we hit, Patty's knee slammed into the steering column, but she never let go of the wheel. Below, my mother ran to help Ben. The boat turned and the wind started hitting us from the side. The sails began flapping wildly. Terror became the engine of my response.

I didn't think, I simply implemented. I did those things necessary to save the boat, to save my family. *Get off this reef,* I thought, *we've got to get off here fast.* If we were caught, the waves slamming into us from the side could break up the boat. I had to regain control of the boat. Our lives depended on it.

I turned on the engine. Miraculously, we floated off the reef. *Mon Pays Bleu* was free, but I had no idea how much damage had been done. I gambled that we would stay afloat, and motored toward deep water. Patty and I fought the sails. We practically dragged them down. That done, I leaped down the companionway and started pulling up floorboards to see how much water we were taking on. That would determine our next move.

The bay was leaking slowly into our lives. A small pool of water was forming in the bottom of the bilge. *Mon Pays Bleu* was leaking, but it was nothing the pump couldn't handle. The hull had held tight.

We'd all suffered minor bumps and scrapes, but no one had been hurt. When we were off the reef, I inspected the boat. I believed we were not in imminent danger of sinking. The only serious damage had been to my confidence. And that had been badly wounded. As soon as it was safe, I went below and turned on our radio. "Mayday," I began, "mayday . . ."

The coast guard monitored our voyage to safe anchorage.

Years later, when once again it seemed like clear sailing ahead of me, this was the lesson I should have remembered.

PART ONE

THE PROMISE

"THE UNITED STATES PATENT OFFICE IS
READY TO GRANT PATENTS FOR MEDICINES,
ALTHOUGH IT IS AN OPEN QUESTION IN
PROFESSIONAL ETHICS WHETHER A PHYSICIAN
SHOULD PATENT A REMEDY . . . ."

—Scientific American, *September 1896*

1 DEATH IN THE COURTROOM CAME SLOWLY, AS A torture, rather than as a sudden and climactic event. Maybe it would have been easier for me if there had been a single moment to which I might point and say, "There, there it is. That's when it happened." But it wasn't like that at all; the verdict was the product of a thousand small decisions by the judge, Roderick McKelvie, each one of them legal, each one of them defensible. At the conclusion of this trial the judge could bang his gavel mightily and declare that justice had been done. Maybe that's what caused me to lose the final remnants of the deep faith I once held in the American legal system.

This was the second trial. The first trial had ended with our complete victory. A jury had found for us on every issue, reaching a unanimous verdict in our favor on 103 different questions. But the judge, McKelvie, made that decision disappear as easily as if he were sponging chalk off a blackboard. It was an act of legal legerdemain of such magnitude that once I would have believed it to be quite impossible. It was as if that trial had never happened. As if there hadn't been a parade of witnesses and thousands of pages of testimony, and motions and objections and rulings, as if I had never sat in that courtroom and fought for the survival of my company.

I had grown up believing that a courtroom wasn't so much a large space as the physical embodiment of a grand concept called equal justice for all. Perhaps the people who worked there every day, the judges and clerks and bailiffs and lawyers, saw it simply as their place of business. But with the exception of fighting a speeding ticket, I had never been inside a courtroom in my life. So when I'd walked into that courtroom in Wilmington, Delaware, for the first time I'd felt a sense of awe. I'd felt the majesty of this idea of justice that had been passed down through the centuries. Corny, absolutely—yet completely true.

3

But as I sat at the counsel table during the second trial, watching Judge McKelvie exercise his powers with the arrogance of a king, I began to understand how wrong I had been. The law wasn't an established doctrine to which the facts of a dispute could be applied; the law was simply whatever the judge said it was. In experimental science, the same set of conditions will produce the same result every time, no matter who conducts the experiment. The laws of nature are immutable; they are not open to interpretation. But the laws of man are easily bent to suit the whims of those who enforce them. As I've learned, the law is a cornucopia of precedent from which a knowledgeable judge can carefully pick bits of case history to determine the outcome of the case in his courtroom. That's why a judge's decision is called an "opinion" rather than a "fact."

To tell the truth, I was embarrassed to be so old yet so naïve about how the law really worked.

The trial had all the trappings of justice being done, all the players played their roles, the institution was treated with respect, the rules were followed. All the attorneys wore somber suits and appropriate ties and spoke the language of the law, but this was just the appearance of justice, no more real than the warmth of the sunlight in a photograph of a beautiful summer's day.

I had come to the legal system as the CEO of a biotech company named CellPro. We were in the business of cell separation and had developed technologies that would help save lives—including, as it turned out, my own. People who might have died were alive because of the discoveries made in our laboratories—and we were very proud of that. At the heart of our system was an antibody, a nearly invisible molecule with the ability to seek out and attach itself to a specific target cell. Another company, a medical-products behemoth named Baxter Healthcare, was also creating a cell-separation system, although Baxter was at least two years behind us in development of a marketable product. Like our Ceprate system, Baxter's Isolex system used an antibody, but it employed a completely different antibody from the one we used.

The question of whether our Ceprate system infringed patents licensed to Baxter by Johns Hopkins University through the medical-products giant Becton Dickinson—where Baxter's antibody had been developed—had been the source of a business dispute between CellPro and Baxter for several years. Although two of the most accomplished patent attorneys in America had determined, after examining all the pertinent documents, that we did not need to license these patents from Baxter, to satisfy any potential legal problems that might arise in the future we had attempted to make a deal with them. In some important ways the two companies were complementary rather than competitive, and both could benefit from a settlement.

We tried very hard to make a deal. But after several years of negotiations we were unable to reach an agreement, and so we ended up law-firm-to-law-firm in the federal district court in Wilmington, Delaware, battling Baxter, Becton Dickinson—another billion-dollar multinational medical-products company—and Johns Hopkins University. David versus three Goliaths. This was simply a business dispute that had gone terribly wrong.

Before joining CellPro I had worked for Baxter for several years, and I knew it to be a litigious company. Baxter had nothing to lose by pursuing this case. The cell-separation program was only a minuscule part of its business. Conversely, we had everything to lose. The only thing we did at CellPro was cell separation. It wasn't just our core business, it was our only business. We did cell separation, and we did it better than any other company in the world. By the time we met Baxter in the courtroom for the second time, we had the only government-approved product on the market and were involved in sixty different clinical trials. Our cell-separation technology was being used routinely to treat leukemia and lymphoma transplant patients. But across the country, clinical researchers were also using our Ceprate device in experimental programs to treat an extraordinary variety of diseases, among them breast cancer, multiple myeloma, childhood leukemia, diabetes, AIDS, and a variety of gene therapy programs, as well as autoimmune diseases including multiple sclerosis, rheumatoid arthritis, and lupus. We had an adaptable platform, and we were just beginning to exploit it.

It seemed beyond my comprehension that a technology with this extraordinary potential could be held hostage in a business dispute, but that was precisely the situation as we began the second trial. I was losing hope that we would win this time; Judge McKelvie had made certain of that during the previous six months. After reinterpreting the meaning of the patent claims to square with Baxter's case, ruling by ruling he eliminated each one of our defenses. Then he declared us guilty of infringing the patents, and empaneled a jury solely to determine the damages we would be forced to pay.

I was just another player in this farce. I knew it, but there was absolutely nothing I could do about it. Our lead attorney, a wise man named Coe Bloomberg, had warned me, "It's going to get a lot worse before it gets better." I didn't know how that could be true. It didn't seem possible to me that the situation could get any worse. So each morning I put on a nice suit and sat quietly at the counsel table, still unwilling to accept that rational people would let this tragedy proceed.

The courtroom itself might have served as a movie set. It was large and imposing, with dark wood paneling covering the walls. There were no windows, which isolated the proceedings from the outside world. I found that to be an appropriate metaphor for the trial.

The judge, Roderick McKelvie, fit perfectly into the surroundings. Tall and graying, he ruled his courtroom like a despot, surveying the proceedings through reading glasses that lent a touch of intellect to his appearance. Earlier in his life, McKelvie told the *National Law Journal,* he had intended to follow his father, an orthopedic surgeon, into medicine. But after observing a bloody knee operation he had changed his mind, explaining, "I liked the idea of driving a Mercedes, of golf on Thursday. But I just didn't like the blood. I didn't understand how my dad could stand to do surgery every day." After 18 years as a practicing attorney, McKelvie had been appointed to the bench by President George Bush in 1992. Perhaps coincidentally, Bush had once served as director of the Central Intelligence Agency—as had McKelvie's ex-father-in-law, Richard Helms. Judge McKelvie was politically well connected, at one time running Delaware's judicial nomination committee for the governor. Prior to the trial we had heard that he was considered the likely choice to fill the next vacant seat on the court of appeals. Clearly an ambitious man.

We had been warned by people who worked in his courtroom that McKelvie had a tendency to make up his mind early in a case and stick rigidly to his decision. I understood that failing only too well. As a CEO, I knew that I was sometimes guilty of the same thing. At times, after hearing only some of the facts, I'd reached a conclusion and then been slow to realize that I might be wrong. I too had a tendency to dismiss facts that failed to support my decision. So I ended up as wrong as the racehorse owner who saw only the solid legs of an animal beneath a fence and ended up owning a mule.

At one point during the trial it occurred to me that Judge McKelvie could not possibly appreciate what was at stake. That realization hit me with the force of a squall at sea. I found that astonishing, but as I thought about it, it was completely understandable and certainly not his fault. To him, to the attorneys, this was a patent law case. Nothing more. It was about property, not people. It was about the application of law, not about lives. Patient care was not an issue. The future of this cutting-edge medical technology wasn't an issue. I'm certain the judge never considered the effect his rulings would have on so many lives. And, I suppose, legally that was proper. But when I remembered all I had been through only months earlier, and the incredible work that had been done by so many people to save my life, it seemed like a terrible shame.

There was nothing more I could do in that courtroom but wait for McKelvie's final decision. I sat politely in the chamber day after day, my hands clasped to white knuckles on the table in front of me, as the anger and bitterness grew inside me. As surely as I had survived a terminal illness, I knew if we lost this case that the technology which had saved me would disappear. I had told anyone who would listen to me that lives were at stake here, but no one had believed me, even though I was living proof of the power of modern medicine. The fact that I was able to be present in that courtroom was a miracle. The fact that I had to be in that courtroom was a tragedy. So as I sat there listening to the judge, McKelvie, excoriate the people who had, against all odds, made possible a treatment that could save the lives of patients suffering from an extraordinary variety of diseases, the only thought in my mind was, what a waste, what an incredible waste.

2 AT HEMASCIENCE, A SHEEP HAD BEEN PATIENT number one. I was number two.

HemaScience was a medical device start-up. I had left a secure job as vice president of marketing at Cytogen, a small company developing technology that would enable monoclonal antibodies to deliver lethal drugs to tumor cells, to be part of it. I had been at Cytogen for less than a year, and I had expected to be there for quite some time. My wife, Patty, and I had just bought a house in New Jersey, and we were just getting ready to move into it, when an entrepreneur named Bill McLaughlin asked me to fly out to Orange County, California, one weekend to look at a blood separation device he and his partners were developing in a garage.

I couldn't resist. A new medical product being built by hand in a garage? That was the stuff of my dreams. It was an opportunity for me to do real hands-on product development, to take a crude prototype and turn it into a company. This was as close as anyone in business gets to being a pioneer. Every young businessman knew the legendary story of Dave Packard and Bill Hewlett, who had started Hewlett-Packard in Packard's garage with $538 in working capital and a dream and built it into a multibillion-dollar electronics conglomerate.

As McLaughlin led me into his associate Bill Miller's garage, I saw two cars, some tools, and a large workbench. Sitting on the workbench was a strange looking device. It was a clear plastic cylinder about ten inches high and five inches wide. A tube ran into it through a port in the top and two tubes came out of the bottom. It was hooked up to a motor with a belt drive. "This is it," Bill said proudly.

"It" was a plasmapheresis device, a membrane separator capable of harvesting only the plasma from a blood donation and returning the remaining components to the donor. Plasma was used to make a variety of drugs and biological products. The difficulty had always been getting rid of the cells and platelets in plasma, which this device accomplished by utilizing a rotating membrane. Normally, the process of separating blood was done in a large centrifuge. It was a time-consuming, cumbersome process. If this machine worked, I knew, it would revolutionize the blood plasma business.

After I'd signed a confidentiality agreement, McLaughin hooked up a bag of blood and turned on the motor. Bright red blood streamed through the tubing into the device and seconds later came out of one of the bottom ports—and through the second port flowed clear plasma. "Wow," I said softly. "That's very impressive."

It probably wasn't as thrilling as discovering the Beatles in a dark club in Liverpool, but to someone who understood plasmapheresis—separating the plasma from whole blood—as I did, it was pretty exciting. "Come and help us start this thing," McLaughlin said.

Patty and I had already moved three times. But as I flew back to the East Coast that Sunday night I tried to figure out the best way to tell her we would be moving one more time. McLaughlin had offered me a small piece of equity in the company. This was our chance to own a piece of the gold mine.

My family never moved into that house in New Jersey. Officially I became the vice president of marketing at HemaScience, but as we only had five employees everyone did whatever had to be done. This was a classic ground level start-up. I had never worked so hard in my life, and I loved every minute of it. In addition to running marketing, I was our purchasing agent as well as an important part of our janitorial staff. I did the budget in the morning, counted tiny transistors and put them in little packets in the afternoon, and swept the floor in the evening.

We began by designing the instrument, practically on the back of matchbook covers. I knew this field cold: I knew what was important to end-users and worked closely with the engineers to give them what they needed. During the design process, for example, the instrument employed four small pumps, and we needed to figure out how to eliminate one of them. I woke up in the middle of one night with the answer and quickly scribbled it down on a sheet of scrap paper. By reversing one pump we could get rid of another one. It allowed us to streamline the console and make the device much more user-friendly. I took my design in the next morning and handed it to Paul Prince, one of our engineers. "This'll work," I said. "It'll solve the problem."

After looking at it carefully, Paul agreed, "Yeah, it will."

Eventually we built a prototype device. One of the very first parts that had to be tested was the pressure monitoring system. The instrument removed the donor's blood then pumped the cellular components back into the donor's vein through the same needle. If the pressure got too high during reinfusion it could severely damage the donor's vein. The pressure monitoring system was supposed to control blood flow. It needed to be tested using real veins. This was potentially a very dangerous test. We decided to use a sheep, because sheep blood is fairly similar to human blood, and sheep were readily accessible. One of our people went to the slaughterhouse and bought a live sheep. We knocked it out with an anaesthetic, found a vein in its neck, and attempted our first *in vivo* run.

It was a moment I will never forget. We huddled around the pressure controller and turned on the instrument. One person was taking pictures. The motors hummed softly. The sheep was lying motionless on the table. We watched silently as blood began flowing through the system. It worked. The system operated within the necessary pressure ranges. Purified plasma flowed out and the cellular components went back into the sheep. The whole circuit worked perfectly. We proved that it would work in an *in vivo* system.

At the end of this experiment the sheep woke up and was absolutely fine. So we sent it back to the slaughterhouse.

We were ready to do our first donor runs with a human being. I called Dr. Gail Rock at the Ottawa Center of the Canadian Red Cross, with whom I had worked while at a different company. Her lab was one of the best in the world for conducting this kind of experiment, and if we did it outside the United States we wouldn't need U.S. government approval. After explaining what we were doing, I told her, "I think we're ready to try a live donor run with this thing. You up for that?"

"I'm ready," she said. "Let's go."

I went to Canada with a very talented engineer named Don Schoendorfer. We were ready to make a test run on a human being—all we needed was the first donor. Everybody else in that room had a job to do: Gail, Don, the head nurse Nancy McCombie. That left one person. "I guess that leaves me," I volunteered. "Let's set it up and run it."

I wasn't the sheep, I was the guinea pig. Two needles had to be stuck in my vein extremely close together; one would be used to draw blood, then reinfuse it, the other to monitor the pressure as blood flowed through it. It was pretty tricky; the needles had to be as close as possible without touching and they couldn't go completely through the vein. I took off a cufflink, neatly rolled up the sleeve of my shirt, and got ready for the test.

I liked being first. I liked the concept of literally being on the cutting edge of science. Doing it made me an active participant rather than a passive observer. And that was important to me. Most advances in medicine are made in small steps rather than great leaps. For safety, the system is made to work that way. Some steps are riskier than others. But there was a person who took each step for the first time. By the time a new drug arrives on the shelves or a shiny new medical device stands in a hospital or in a doctor's office, numerous people have taken risks to make sure it is safe and does precisely what it claims to do. Sometimes people die during that process; few things work perfectly the first time every time. Drugs may have unexpected side effects, devices can break. Death, too, is an inherent aspect of progress.

I had complete confidence that this device was safe and it would work exactly as designed. I knew every part of it, every valve, every clip. I'd helped design it in my sleep. I'd helped put it together. If I felt there was any danger I might have hesitated. I knew it was safe but, still, when Don leaned over to start it . . .

It was impossible not to feel some apprehension. Knowing myself, I probably made some joke as we waited. Maybe something like, "Gee, if the first thing I say after I get my blood back is 'Baaahhhh,' then we know we've got a problem." I'm confident that outwardly I showed little sign of being nervous.

Within seconds it was obvious the system was safe. We ran the test and got beautiful data, just beautiful. This was a huge achievement. We had a compact membrane-based system that successfully separated plasma from whole blood and returned the remaining components to the donor. There was a lot more to do before it was perfect, but it worked!

Soon afterward we began clinical trials and eventually received FDA approval. We'd taken the instrument from a prototype in a garage to a sophisticated medical system, which included a highly automated instrument and disposable package, and we'd done it for about $5 million. It was a remarkable accomplishment.

Eventually two medical products giants, American Hospital Supply and Baxter, wanted to buy HemaScience. Almost all successful start-ups end up being acquired by larger companies. It is a very desirable exit strategy for the original investors. I thought AHS was a better fit for us, and we expected to conclude a deal with them. Based on the fact that we had an approved device and the potential to develop further products based on the same platform technology, we believed the company was worth a minimum of $50 million. And those were the kind of numbers AHS was whispering to us.

But in the middle of our negotiations Baxter acquired American Hospital Supply. With one great multibillion-dollar gulp our options had become severely limited. So we began negotiating with Baxter. This was the real beginning of my education.

▷ I've spent my entire professional life in the business of science. I've been a blood man. I'm not sure where my interest in science came from, but I've always been fascinated by biological systems. I was born with a curiosity about life in all of its forms. Perhaps I inherited that from my father, who taught me to love the outdoors. Almost as long as I can remember my younger brother and I were encouraged to discover nature, particularly the kind of nature found in trout streams. We lived in northern California, and my family would drive into the mountains to camp near remote stretches of the river for much of the summer. Often during those times I'd wander away on my own trips of discovery. I was always comfortable with nature.

Science gave me the tools I needed to satisfy my curiosity. For a time I thought I might be a marine biologist. I learned to scuba dive because it allowed me access to the extraordinary variety of creatures living in the ocean. I've seen spectacular things in the water—I've seen a lobster with antenna at least six feet long; I've seen sea anemone with trunks as thick as small trees; I've seen a giant Pacific octopus; and I've dived with sharks in the Caribbean. One day while free diving off the beach in Monterrey Bay I was floating on the surface when a giant manta ray at least 20 feet across lifted gracefully off of the sandy bottom directly below me and gently glided away.

At UC Berkeley I majored in zoology, spending weekends at their marine laboratory collecting specimens from tide pools, then trying to identify and categorize the species I found there. I spent hundreds of hours staring into a microscope without ever being bored. Even today I can look at a drop of water under a microscope and identify a dozen types of protozoa.

I didn't particularly enjoy chemistry or physics, but biology . . . biology always made sense to me. I found a logic in biology that appealed to me. I was continually conducting experiments to try to understand how life works just a little more clearly.

To the embarrassment of my parents, who had sent me to a strict Seventh-Day Adventist high school, at UC Berkeley I'd become a long-haired hippie. My mother wanted to disown me, my father had a difficult time with my 1960s rebellion. We didn't agree about anything, a situation we resolved by rarely speaking to one another.

I didn't know what I wanted to do when I graduated from college. Fortunately, I'd graduated right into the biotech revolution. Scientists were beginning to examine life in terms of biomolecular components. So I cut my hair, put on my suit, and went to an employment agency. A counselor there decided I would make a good salesman for a medical products company. The only thing that might hold me back, he pointed out, was that I had absolutely no sales experience.

A salesman? The only thing I knew about salesmen was that they came in one hair length—short: The last thing I wanted to be was a salesman. But the more I thought about it, the more sense it made. I was a very hard worker, I could speak state-of-the-art science, and I had always been able to motivate people. Selling was probably a good way to survey the field and decide exactly what I wanted to do—and earn some money while I was doing it. Eventually I was offered a sales job by OrthoDiagnostics, a Johnson & Johnson subsidiary that produced blood reagents and diagnostic tests. These were the antibodies used by blood banks and hospitals to determine blood type and blood-type compatability for transfusions as well as some types of therapy.

I knew very little about blood when I accepted the job. Ortho put me through an intensive three-week-long training course. Much of it was an extension of what I'd learned in school, so it was not difficult. But it provided the basis of knowledge on which I built my career.

Ironically, the blood system is the closest thing to the ocean that exists in nature. Maybe that's why it so intrigued me. If you believe the evolutionary biologists, the blood system is an internal ocean—it has the same pH or acidity, the same salinity, several of the same characteristics, and the same buffering capacity of an ocean. Experience has also made me believe that it responds to the phases of the moon, just like ocean tides. Police will confirm that the full moon really does have an effect on behavior. As I discovered in these classes, the human blood system is an incredibly complex and endlessly fascinating environment.

Ancient man recognized blood as being the life force 20,000 years ago. The early Egyptians and Greeks understood that blood had the power to control health and disease. Although people once believed that personality could be improved through blood transfusions, scientific research into the components of blood and their functions began in the seventeenth century. But only in the last few decades have we finally begun to understand the complexity of the blood system and how it can be manipulated to maintain health and even cure many of the most devastating diseases.

The blood circulatory system really is the river of life; it's a closed system in which all bodily commerce takes place. The exchange of gases—oxygen and carbon dioxide—takes place in the blood system, waste products are eliminated through the blood system, cells get the nutrients they require through the blood system, and our entire immune system operates through the blood system. Unfortunately, disease is also spread through the blood system. Blood consists of two basic components: cells and plasma. If you take a bag of blood and let it settle for about an hour you'll see the basic structure of blood. The top half will be the plasma, the fluid portion of the blood, the bottom half will be the red blood cells. These layers will be separated by a very thin layer of white blood cells known as the buffy coat—but that thin layer is the entire immune system.

After three weeks with Ortho I thought I knew a lot about the blood system. But I had not even begun to appreciate the wonder of it. And in my wildest dreams, I could never have imagined that one day my life would depend on the ability of scientists working a few doors down the hall from my office to manipulate my blood.

To my great surprise, at Ortho I discovered that at heart I am probably a salesman. That was not something I would have suspected about myself. But I found that I enjoyed rolling up my sleeves and working with doctors and scientists to understand what they needed and how I could help them fulfill that need. I became as much a product consultant as the man who filled the orders. Sales appealed to my competitive nature. I love to win, I always have; I love to dive the deepest, produce the best product, sell the most. The bottom line is an inflexible measuring stick. You either do it or you don't. Fortunately, I did it; I became one of Ortho's top salesmen.

One morning in 1972 I had an appointment at Puget Sound General Hospital in Tacoma, Washington. As I walked in I noticed a very attractive young woman, a medical technologist, standing behind the counter. We spoke briefly; she was very businesslike, sort of huffy and kind of abrupt. I thought she was beautiful. She was petite and full of spunk and energy and fire. It also seemed to me that for some reason I was making her nervous. I wasn't sure why; I figured she was a small-town girl and I appeared to be a pretty slick salesman.

I began to look forward to my trips to Tacoma. The only problem in our burgeoning relationship is that she didn't seem to notice we were having a relationship. She seemed completely uninterested in me. Imagine Meg Ryan or Sandra Bullock as the feisty, reluctant object of affection resisting the charms of the earnest young salesman—and that was Patty and I.

More than a year later we met accidentally at a convention of medical technologists in Yakima, Washington. She was studying for an advanced certification in blood banking, she told me. Blood banking, the storing and processing of blood, was an area I knew very well, and I offered to help her. On our first date we sat in a park on a hill overlooking Lake Washington. It was a lovely day. We sat there discussing blood banking methodologies, especially the different antigen systems and the products Ortho had that dealt with them. The closest we got to romance was determining if a person with blood type A married a person with blood type O, what blood type would their offspring not be.

We discovered that we had much in common—in addition to our interest in blood. We came from generally similar backgrounds, our value systems were compatible, we enjoyed science and we even liked the same music. Thank you, Elton John. We were just very comfortable together. I had never met anyone quite like her. I could feel myself falling in love with her.

I'd never been shy around women, but Patty made me feel shy.

I'd never met anyone even remotely like her. I still haven't. She brings her own perceptions to every moment—although admittedly there are times when I have absolutely no idea what she's talking about. She will make up her own words to describe things. When we were dating, I remember, we went shopping for a television set. After examining a small set for a minute she decided against it, telling me, "You're in a situation where instead of you watching it, it's watching you." The scary thing was that at that point I understood exactly what she meant.

We had been dating about a year, with absolutely no discussion of the future, when Ortho promoted me to product manager of the blood bank reagents segment of their business. I would be running a major part of a division doing $50 million a year. The problem was that I had to move to their corporate headquarters in New Jersey. Patty had never been to the East Coast, but she was ready to leave the Northwest. For someone who had not traveled extensively, Patty was open to change.

Neither Patty nor I remember how I asked her to marry me. I'm sure it was more romantic than "How would you like to go to New Jersey?" But everything happened in a great whirlwind: the job offer, the decision to move to the East Coast, our marriage. We were married in a small church in Tacoma in 1974 and immediately moved to New Jersey.

Ortho was a risk-adverse, conservative company. I'd been the hot-shot young guy in the field, a smooth talker, aggressive and creative. But while that served me well as a salesman, my personality worked against me in the home office. Ortho had become successful doing business in a traditional way, and they intended to continue doing business that way until they weren't successful. I was told what advertising agency I had to use, how to prepare my business plan, how to sell, how to market. There was little room for innovation. I remember so well sitting in my oak-paneled office and wondering, *What am I doing here?*

I liked many of the people with whom I worked, but getting even minor tasks accomplished often required me to slog through several layers of management. At heart, I wasn't really an Ortho kind of guy. I could see that the climb up that particular corporate ladder was going to be long and ugly. All around me I could see the foundations of the biotech industry being built. I wanted to be part of it. I wanted to build something with my name on it. So when the executive placement services, the head-hunters, began calling, I listened.

After eighteen months as an Ortho executive I accepted a job as the first product manager at the Haemonetics Corporation in Natick, Massachusetts. Haemonetics manufactured medical equipment that separated whole blood into its components, as well as pioneering interoperative blood salvage, bloodless surgery, in which a patient's own blood is collected from tissue around a wound, washed and returned to his body.

It was at Haemonetics that I realized there was a significant difference between the diagnostic products we'd been selling at Ortho and medical devices, the instruments and equipment that actually interact with patients. The day I learned that was the day my life was changed forever.

I was at the Central Blood Bank of Pittsburgh, field testing a new plasmapheresis system with the young, aggressive medical director, a doctor named Ron Gilcher. I set it up and we made a couple of donor runs and it ran very well. I was pleased with how well the system had performed. Gilcher was very impressed. "You know," he said, "this works pretty well on donor plasmapheresis, but what we really ought to be doing is using it on patients."

Use it to treat patients? I'd never even considered that possibility. Patients weren't really part of my job. Besides, this device hadn't been designed to treat patients.

Gilcher didn't care what it had been designed to do; he knew what he wanted to use it for. "Let's go," he said.

Go? Go where? Less than an hour later I found myself wheeling this device into the intensive care unit. A woman suffering from an autoimmune disease was having an acute attack. Her own antibodies were attacking her tissue. She was very sick. "We're gonna do a plasma exchange on this patient," Gilcher explained. "This'll really help her."

I was scared to death. I knew that if my boss back at Haemonetics knew I was using this device to treat a patient he would have stopped me. Stopped me? He would have gone crazy. But I was caught up in Ron Gilcher's enthusiasm. Because the device was not intended to be used on patients, we had to do some pretty fast jury-rigging. I had to make important decisions right there. The patient was lying in bed, right in front of me, and she was suffering. My God, I thought to myself, what am I doing here? This isn't what I'm supposed to be doing.

I wasn't even confident the device would hold up. I was afraid it might blow up right in the middle of the process. But it didn't, it worked beautifully. We exchanged about four liters of plasma in this patient. We eliminated a substantial number of the antibodies that were attacking her tissues and substituted saline solution. For me, this was an amazing experience. I was fighting on the front lines of medicine—and I loved it. It was exhilarating.

Hours later my heart was still pumping adrenaline. I couldn't believe what we had done. The next day I went to see the patient and she was significantly better. Her crisis was over. She was awake and smiling. I think this was probably the first time I really understood that the devices we were developing and marketing actually made an impact on people's lives. Until that day I had been completely disconnected from the end-user, the patient. But I would never forget it.

I don't think I ever again thought of myself as simply a product manager or a salesman. I was part of the world of medicine, helping to provide the tools that physicians wanted and needed to better serve their patients. There was great satisfaction in that.

Haemonetics was a fine company, but it was a very small company, almost a medical mom-and-pop operation. After spending five years there it was obvious I was pretty much tapped out in terms of growth within the company.

The blood business had changed completely since I'd started at Ortho. At that time blood technology had been pretty much limited to identifying the numerous blood group antigens and components. We could pump it out and pump it back in. But for the first time scientists were beginning to discover ways of actually using blood products to fight diseases of the immune system and various forms of cancer. The business of blood was changing gradually from diagnostics, what's wrong with you, to therapeutic, a means to treat it. I knew enough about science to be really excited about all this, and I wanted to be part of it.

I became the vice president of marketing at the Cytogen Corporation. Cytogen was among the first companies attempting to use monoclonal antibodies as submicroscopic delivery trucks. For example, binding imaging agents—materials designed to show up on X-rays—to monoclonal antibodies that would seek out tumor cells made it possible to find out where in the body these tumor cells were growing. The dream was to be able to deliver lethal drugs directly to the tumor.

By the time I started working at Cytogen both of my sons, Jamie and Ben, had been born. But we had moved so often that Ben, when asked where he's from, still responds, "I'm from all over." Through it all Patty was amazing. With each move she had to uproot our growing family and settle someplace new. New Jersey, Massachusetts, back to New Jersey: with each move came a new house, new schools, and new friends. She never complained about it. Truthfully, I often worked such long hours that she had to do much of it herself. She gave up doing many of the things she loved to support my career and raise our boys. She's a talented artist, for example, but rarely found time to paint. Through all the years, particularly when I was doing a tremendous amount of traveling, Patty never missed one of the boys' Little League or soccer games or swim meets or teacher's conferences. Whatever the need, she found the way to fulfill it.

When I arrived at Cytogen I'd already worked at established companies, and at early-stage companies I'd been in sales, marketing, and management; I'd learned the technology from the classroom to the patient. About the only thing I hadn't done was be involved in a start-up, a place where an idea was turned into a company. That really appealed to me. The availability of venture capital had made start-ups the center of the action in the biotech industry. So when Bill McLaughlin contacted me about going to HemaScience, I was ready to make a move.

Five years later HemaScience had a proven product worth a small fortune and we were negotiating its sale to Baxter Healthcare. Ironically, before being purchased by Baxter, American Hospital Supply had bought Haemonetics, a major competitor in this area and my former employer. Theoretically, this created anti-trust problems for Baxter—if they bought us they would gain complete control of this market. Baxter told us they had to suspend negotiations with us until the acquisition of AHS was complete and Haemonetics had been divested.

I believed them.

That put us in a very difficult position. We were running out of money, but other potential buyers knew we were in discussions with Baxter and didn't want to compete with them. So while we were waiting to resume negotiations, Baxter loaned us several million dollars to continue our development and marketing efforts. What I should have realized, and did not, was that they were putting us in a stranglehold.

When negotiations finally resumed we had a very difficult time concluding a deal. Baxter's negotiating strategy seemed to be to make a proposal; back away from it when we came close to an agreement; then return with a much less attractive offer. Negotiating with Baxter was like being the greyhound chasing a rabbit that could never be caught. At one point when we were close to reaching a deal the person with whom we'd been negotiating was suddenly replaced, forcing us to begin all over again. As time passed and we got more in debt to Baxter the deal got progressively worse. We finally ran out of options, and we had to pretty much take whatever deal they put on the table.

Baxter believed it had made a fair deal; I thought it was predatory. We accepted what Baxter called an "earn-out deal." They paid $25 million up front, which went to our shareholders, with the potential of making an additional $25 million from profits earned over the first five years of the deal.

After we agreed to that, Baxter added that we had to pay back the $9 million we had borrowed out of the proceeds—although they did not charge any interest.

Baxter then sacrificed potential profit to build market share, knowing that as part of the purchase price a high percentage of any profits earned during that period would be shared with HemaScience stockholders. They distributed hundreds of our $15,000 plasmapheresis instruments for free, then cut the price of the disposables—the replaceable parts used during each donation—to the bone. In marketing this is known as very aggressive pricing. They basically gave away the product to get into blood centers and hospitals and front-loaded significant expenses, so by the time the device became profitable the earn-out period had expired.

Almost ten years later, when it was clear there was no significant distribution, HemaScience shareholders filed suit against Baxter.

Officially, we were acquired by Fenwal, a division of Baxter. Fenwal is an industry leader in the manufacture and distribution of blood collection, processing, storage, and transfusion systems. Apparently I'd made an impression on Baxter's executives during negotiations, because I was offered an incentive-based contract to serve as Vice-President of Marketing for Fenwal Automated Systems. It was too good a deal to turn down. In addition to continuing to market our plasmapheresis system, I was given responsibility for all their automated blood processing systems, including one of their most important products, the CS-3000 blood-cell separator.

Had I not been through the experience of a start-up, I might not have accepted their offer. But after five years of living close to the street I was ready for a change. I knew I wasn't going to stay at Baxter for the rest of my life, but after fighting the storms of a start-up it was like pausing to rest on a sunny beach.

As it turned out, Baxter was never a comfortable fit for me. It was everything I disliked about working at a large corporation—but bigger. It was an extremely political organization. There were layers and layers of bureaucracy, which caused development to move with the speed of a glacier, but gave executives a lot of insulation from real responsibility. After managing the development and launch of a new product at a small aggressive entrepreneurial company, I was in total corporate shock.

Baxter's strength was its marketing muscle. Baxter executives were almost fanatic in their desperation to dominate the blood bank business. They had to have 100 percent of the business; ninety-seven percent was not acceptable. At meetings, market share was almost always the prime topic of conversation. With the exception of HemaScience, Baxter was certainly not a leader in price-cutting; I was always told not to cut prices, that our business strategy was to beat our competitors by out-marketing them with quality products. In selective cases, however, it seemed to me that they lowered prices to gain market share, then after they had acquired the business they would begin raising prices.

They did nothing illegal, but they always pushed the envelope. I remember very clearly that at one point a senior executive whom I liked and respected a great deal told me that Baxter had conducted a survey of its managers. Basically, he explained, the managers felt that Baxter was a strong company with good leadership, but that these same managers had given the company low marks on ethics. The employees thought the company had an ethics problem. This executive told me that the company needed to do something to improve that perception. After working there, I agreed with that assessment.

I discovered quickly that several Baxter managers resented me. They were Baxter lifers; they'd worked their entire careers to reach the executive level. I was younger than most of them and had parachuted in from a small start-up to a senior level position. Some of these people reported to me, and it was clear they didn't like it. They really made me feel like an outsider.

After three years at Baxter I was pretty much washed out. My batteries had been recharged and I was ready for a new adventure. I was ready to look for another start-up situation. The breaking point had come at a meeting at which I had strongly supported modifying one of our products. This innovation would have changed the way blood transfusions were done. The engineers had worked out the details. It worked, it was cost-efficient and it greatly improved the existing product. There were a lot of people in the room as we made our presentation. Tim Anderson, the president of the Fenwal division, sat in the back like a judge. At the end of the meeting he decided we weren't going to do it. He gave no explanation—he just didn't want to do it.

I just threw up my hands and decided it was time to leave. Anderson must have realized I felt that way, because Baxter offered me the job as vice-president of its European operations. It was a very exciting prospect, a chance to live in Brussels, Belgium, and travel around the continent. Patty wanted to do it, and we thought it would be a wonderful experience for the boys, so we accepted the job.

It was the right decision. At Fenwal Europe we greatly increased our sales and profits while launching some exciting new products. But after two years there my contract was expiring and we had to make a decision. Jamie was preparing to enter high school and I felt very strongly that the boys needed to go through four years of high school in one place. So we had to decide whether we would commit to staying in Europe for six more years until both boys graduated or return to the States.

Patty preferred to stay. As the boys had grown independent she had started doing things for herself that she'd been neglecting for years. She'd begun painting again and traveling, and visiting the great museums and art galleries of Europe. There was so much more she wanted to do there. Ben enjoyed it too, Jamie less so. I wasn't sure what to do. Baxter offered to extend my contract, but I knew I didn't want to be a lifer there. Eventually I wanted to go back to the world of start-ups. That's where I'd had the most fun. Coincidentally, and completely unexpectedly, I received a letter from a headhunter in Boston named Nathan Berry. I never knew how he found me. He represented a biotech medical device company, his letter explained, searching for a vice-president of sales, marketing, and business development. As I read the letter it seemed as if someone had read my résumé and then designed a job specifically for me. It was difficult to believe that while I was making a career decision a letter would arrive from someone I'd never heard of describing the perfect job for me.

Of course, that was only the beginning of fate's plans for me.

When I showed the letter to Patty she said, "This is you."

I had been scheduled to attend quarterly Baxter meetings in California. En route, I met Berry in a lounge at Kennedy Airport for three hours. The job he described really did seem to fit me like a well-tailored suit. The company was named CellPro, he told me; it was a biotech start-up developing a cell selection and separation technology. Incredibly, it was located just outside Seattle, Washington. Literally and figuratively, it was a home run.

CellPro was a product of the biotech revolution, created to make money by saving lives. It was located in Seattle, Washington, because that's where Dr. E. Donnell Thomas did the pioneering work in bone marrow transplants that earned him the Nobel Prize.

▷ The technology that years later would give birth to CellPro was created on the morning of August 6, 1945. At precisely 8:15 A.M. the *Enola Gay* dropped the only bomb it carried on the Japanese city of Hiroshima. Ground zero was the Aioi Bridge over the Ota River. Little Boy, as this bomb was known, weighed 8,900 pounds. At 1,980 feet above the bridge a bullet-shaped piece of uranium was fired into a larger piece of uranium at a speed of about a millionth of a second, splitting an atom and initiating a chain reaction resulting in a ten-kiloton atomic explosion. By 8:16 that morning 66,000 people were dead and 90 percent of the city was destroyed.

However, this was also the beginning of another chain reaction, one that would continue for decades and save countless thousands of lives. Weeks after that attack and after the similar bombing of Nagasaki that ended World War II, large numbers of survivors of the initial bomb blast began dying for some unknown reason. Medical researchers discovered that their immune systems had been destroyed as completely as the buildings on the ground and did not regenerate, leaving these victims susceptible to deadly disease and infection. These people were dying from the kind of internal attacks that a functioning immune system would have easily fought off.

While it had been known previously that exposure to massive doses of radiation—as was released when atoms split—killed blood cells, experiments conducted on lab animals proved that radiation also destroyed the regenerative cells that would have matured and replaced those blood cells. It was like pulling the plug on an assembly line.

There had been some rudimentary research done in this area. As early as 1922 a Danish scientist had proved that if the bone marrow, the site where blood cells are formed, was protected, the immune system could recover from the devastating effects of radiation poisoning. Something in the soft core of bones caused the immune system to grow, but scientists were unable to identify and isolate that seed.

In 1951 researchers had found that by transplanting a sample of bone marrow or spleen cells from a genetically identical donor they could regenerate the immune system in a mouse. Within weeks of the transplant the immune system would be functioning perfectly—as long as the donor was genetically identical. Just like putting the plug back in the socket.

But the donor mouse had to be genetically identical. British scientists found that they could successfully reconstitute a destroyed immune system using bone marrow from a nonidentical mouse, but within three months the recipient would die from what they termed "secondary disease." As it turned out, secondary disease was caused by the remnants of the donor immune system, T-cells, attacking and destroying the recipient's tissues. Named graft-versus-host disease, GvHD was extremely painful and almost invariably fatal. And until it could be overcome, bone marrow transplants would remain extremely perilous.

In 1956 Don Thomas performed the first human bone marrow transplant, grafting healthy marrow from a donor into his twin brother, who was suffering from leukemia. The transplant was successful and the recipient accepted the marrow, but it was too late; the disease had progressed too far and the patient died.

Eventually Dr. Thomas proved that transplanting healthy bone marrow from closely matched tissue donors could save the lives of terminally ill leukemia patients. The problem he couldn't solve was how to combat deadly graft-versus-host disease. Transplant patients were literally being killed by the donors' immune cells. The odds of finding an unrelated matched donor for a specific patient were about one in twenty thousand, but without a match the patient had little hope of survival. Dr. Thomas increased the odds of survival slightly by using controlled radiation to suppress the immune system, but until a reliable means could be found to fight GvHD, bone marrow transplants—as well as whole organ transplants—just would not be a viable treatment.

The real mystery remained: exactly what was it in bone marrow that gave birth to a new immune system?

Bone marrow is a thick gelatinous factory of blood cells; it's the assembly line that continually pumps millions of new red and white bloods cells into the circulatory system. But when a transplant patient received whole bone marrow, for better or worse he was getting everything made in the entire factory—including T-cells and, in many cases, tumor cells.

For years scientists had speculated that among the complex machinery inside that factory there must be a single cell from which all other types of blood cells eventually developed. A "seed" cell from which would grow an entire immune system and all the blood cells necessary to circulate oxygen to the tissues. Theoretically, this "pluripotent," or all-powerful, cell would divide and differentiate into all of the various types of blood and immune cells found in the body. No one even knew if such a cell existed, but if it did, and if it could be isolated, it would be sufficient to transplant just a few of these cells rather than using whole marrow. Instead of a couple of quarts of bone marrow, a pinhead of these mother cells would give birth to new, healthy blood and immune systems.

Scientists referred to this as the stem cell, and the search for the existence of the stem cell became the Holy Grail of scientists and transplanters. If the stem cell existed, and if it was capable of regenerating the entire immune system, transplants could be made safe and reliable. But that was a big "if." Like treasure hunters searching mountain caves for a legendary gold mine, scientists pursued even the smallest clue that might lead them to a stem cell.

In 1961 two Canadian scientists provided strong evidence that such a cell existed. After destroying the immune and blood-producing systems of mice by exposing them to radiation, they transplanted bone marrow from a genetically identical strain. Within weeks single cells in the spleen had divided and multiplied to produce large colonies of blood cells. But proving stem cells existed and finding a method to isolate and purify them so that they might be used in transplants were two completely different problems. It would be almost two decades before stem cells were successfully collected.

In his laboratory at the prestigious Johns Hopkins University Oncology Center in the early 1980s a young doctor named Curt Civin wrestled with the problem. Most prior efforts to isolate stem cells had used negative selection: if all the other types of cells in a blood sample were picked out, theoretically all that would be left were stem cells. But Civin believed that he could create a monoclonal antibody, a biological homing pigeon that is attracted to only one type of cell, that would bind only to stem cells. This was a far more difficult task than finding a specific needle in a haystack of similar needles; it was like trying to create a magnet that would attract one needle and only that one needle—without knowing very much about that needle. And it would be a very tall haystack, too— in bone marrow only one of every 10,000 to 100,000 cells is a stem cell. In circulating blood stem cells are literally about one in a million.

Civin eventually succeeded in creating an antibody, which he named My-10, and planted his flag on a stem cell. But Civin was not the only scientist to successfully isolate stem cells. He may not even have been first. At about the same time Civin was isolating My-10 at Johns Hopkins, Professor Robert Tindle, working independently in Glasgow, Scotland, discovered a completely different monoclonal antibody that bound to blood stem cells. Eventually other researchers discovered more than two dozen different antibodies that bound to stem cells.

One of those people was Irwin Bernstein. In 1986 at the Fred Hutchinson Cancer Research Center in Seattle, Washington, Dr. Bernstein and his associates discovered a monoclonal antibody that attached to stem cells. They named it 12.8. There were several important differences between this antibody and Civin's My-10: in addition to the fact that Bernstein's antibody bound to the stem cell at a different site, what made it far more valuable than Civin's was that it had great laboratory utility. Bernstein's antibody would bind to baboon stem cells, which My-10 did not, meaning 12.8 could be used in experiments with primates. In addition, a biological glue, a B-vitamin named biotin, could be attached to Bernstein's antibody but not to Civin's.

It was another smart young scientist at the Hutch, Dr. Ronald Berenson, who invented a practical method for picking stem cells out of bone marrow or circulating blood. Eventually I would get to know Ron Berenson very well, as we spent a great amount of time together trying to build CellPro. Berenson was a fascinating man on many different levels. He was an extremely intense, dedicated scientist who had become interested in cancer research after the disease had killed his father. Ron Berenson and his twin brother, Jim, who also became an important clinical oncologist, had grown up with a love of science. When they were only fourteen they'd attracted the attention of astrophysicist Carl Sagan by conducting rudimentary experiments designed to determine what forms of life might survive on other planets. In their Portland, Oregon, home they simulated the atmospheres of distant planets by carefully mixing gases, then attempted to grow a variety of microorganisms. In 1970 the Berenson brothers went to Cornell to work with Sagan on this "origin of life stuff," as Ron described it.

Eventually Ron realized that he was a lot more interested in applicable science than theoretical research and decided to go into medicine. Rather than investigating the mysteries of outer space he studied one of the most baffling mysteries on earth: what makes cells suddenly go crazy? Why do they change and grow out of control, eventually overwhelming and destroying their environment, the human body?

After joining the Hutch in 1984 Berenson began working with Irwin Bernstein. Both Civin and Bernstein had successfully found antibodies that targeted stem cells, but until a method could be found to separate these cells from all other cells, that discovery had about as much value as a gold mine on the moon.

Both men began searching for a means of picking out stem cells and only stem cells from bone marrow. It was like trying to find a simple and reliable method for picking out only the A's in a huge vat of alphabet soup. Berenson's technique involved creating compounds. In some ways biochemistry is like playing with a set of molecular Legos. Researchers use the ability of molecules to bind to certain other molecules to create new compounds. Berenson had discovered that Bernstein's antibody could be chemically bound to biotin, a B vitamin which is found in every living cell. What made that so important to him was that biotin has an affinity to a protein named avidin. It's actually more than just an affinity, it's an obsession.

Avidin is found primarily in egg white, where it prevents bacteria from growing and destroying the embryonic chicken and the egg. More than half a century ago it was known that avidin and biotin bind to each other more tightly than a rock star and a supermodel. They don't just stick together, they have a tremendous attraction to each other. In a beaker full of molecules they will seek out and find each other. And when they find each other they don't let go. The bond between avidin and biotin is the strongest such affinity in nature. It is almost as strong as super glue.

Berenson used this affinity between avidin and biotin as the basis of his stem cell selection system. After binding biotin to Bernstein's 12.8 antibody, this "biotinated antibody" was put into a bag of bone marrow. Inside that bag, 12.8 found its intended partners—stem cells—and bound to them. As a result of this first step, a stem cell and a biotin-coated antibody were locked tightly together.

Berenson then passed this mixture through a plastic column filled with tiny avidin-coated plastic beads. The biotin stuck to the avidin as if it were flypaper, carrying with it the antibody and the stem cell. All the other cells flowed through the column into a waste bag at the bottom. At the conclusion of this second step, the biotinated 12.8 antibody and the stem cell were practically glued to the avidin-covered beads.

The problem became releasing the stem cells from this compound. In *his* experiments, Civin had employed a harsh chemical releasing agent, which could damage the stem cells. But by using a turkey baster to stir the mixture, Berenson discovered that stem cells could be released from this embrace in perfect condition by gentle agitation. Most of the biotinated antibody stayed behind, bound to the avidin beads. The result was a purified suspension of stem cells.

The technology worked. There were countless questions to be answered, and it would take many years, thousands of experiments and cost millions of dollars to find those answers, but the technology worked. A mechanical instrument to automate the process had to be designed and tested and retested—it just wasn't practical to depend on one guy with a turkey baster—but all that could be done. Berenson had figured out how to safely separate and collect purified stem cells.

Berenson and Tony Goffe, a talented medical device engineer, eventually designed the instrument, which they named the Ceprate—because it separated cells—that successfully automated this process. Only a few years earlier Berenson and the Hutch might have announced their success with a scholarly article in an important scientific journal, sharing details with other scientists that would enable them to perform additional experiments and extend this chain of knowledge. But in the 1970s a series of events had revolutionized medical science.

Until that time most medical research had been done by people curious to unravel the mechanisms of life and death, people more interested in expanding knowledge and saving lives than in earning a fortune.

Dr. Jonas Salk, for example, gave up millions of dollars by refusing to patent his polio vaccine, instead insisting that it be disseminated as widely and quickly as possible. Scientists didn't earn large salaries or drive sports cars. Most medical research was being done in university labs, supported by government grants, or at the giant pharmaceutical companies.

That began changing in 1973, when Herbert Boyer at USC and Stanley Cohen at Stanford put a gene from an African clawed toad inside a bacterium, and that bacterium began producing protein. For the first time in history, DNA, the basic structure of life, had been removed from its natural state and made to reproduce in another living organism. Boyer and Cohen had made possible everything from the production of new drugs to the creation of a Frankenstein monster. This was the birth of genetic engineering—and the biotech industry.

Recombinant DNA technology, as it became known, showed scientists how to splice strands of DNA into desired segments, recombine those segments with other strands to create new DNA sequences, then produce billions of copies. DNA was the recipe of life, and Boyer and Cohen had figured out how to sweeten the pot. Their technique allowed researchers to mass-produce hormones and other biological products that until then could be found only in the body. It allowed scientists to create new biological compounds. It allowed scientists to begin fine-tuning nature. The biotech boom had begun.

Biological science had never been run as a business. Profits and losses had never been part of the equation. Results were measured in scientific advances rather than dollars and cents. There had been no investors to please, no shareholders demanding increased profits. There had been no annual reports and corporate lawyers, no public offerings, no mergers and acquisitions

While Boyer and Cohen made history, making a profit was something else entirely. Scientists still hadn't begun thinking that way. In 1975, for example, Cesar Millstein and Georges Kohler invented a technique for making monoclonal antibodies, those biological homing pigeons, for which they would eventually win the Nobel Prize. Had they chosen to patent this invention, it eventually might have been worth billions of dollars—but they didn't. They published the results of their experiments and shared their knowledge with the scientific community. Other scientists immediately began using this method in their own work. This was the technique Curt Civin used to make his stem cell antibody. That's the way science had always been conducted.

But those days were done. Within five years the world of biotechnology had changed forever. In 1976, twenty-nine-year-old venture capitalist Robert Swanson convinced Cohen and Boyer to market their technology. In 1980, two years before Genentech, the company they formed, licensed its first product, human insulin, the company went public with a $35-a-share offering.

In less than an hour the shares were trading at $88. Genentech was among the fastest run-ups in Wall Street history. Suddenly, there was a fortune to be found in test tubes. Venture capitalists descended on laboratories, desperate to fund the next Genentech, willing to finance just about any discovery that might lead to a marketable product. It was raining money. As long as it was biotech the venture capitalists were thrilled to listen. And the scientists didn't object. There was no reason a scientist couldn't pursue worthwhile projects *and* drive a sports car.

All of a sudden biology was cool. Scientists had opened the door to life. It was just a crack, one just wide enough to peek inside and behold some of the wonders, but it was more than enough to turn science into a business.

I was just beginning my career as a medical products salesman when all this was happening. Everyone understood that we were witnessing the dawn of a new age of scientific discovery; I'm not sure anybody realized how completely economics was going to shape the future of the medical science industry. I remember being extremely excited about these discoveries that were turning a traditional, stodgy industry into an exciting place to be working. At times it felt as if we had suddenly become the center of the scientific universe.

It was a good time to be talking monoclonal antibodies. That term engendered the same kind of dreams that words like "Internet" and "software" would produce years later. Biotech was a hot area, and Ron Berenson had a biotech product that worked in the laboratory. The initial scientific work had been done. Now it was time to get down to business.

After graduating from Stanford and Yale medical school, Berenson had returned to Stanford to do research. There he had worked under the direction of a small, compact man seemingly bursting with energy named

Rich Miller. Miller had subsequently left academia and worked with venture capitalists to successfully found two biotech companies. But he and Berenson had remained in close contact. Both men recognized the potential value of a viable stem cell separation system.

Berenson met with Rich Miller in Miller's office at a company he had helped found, IDEC—today a $2 billion company—to review the data. Miller was enthralled. After listening to Berenson for more than two hours Miller told him, "It's terrific. The system works, it's feasible and there's a clinical need." In Miller's office that day CellPro was founded.

After licensing rights to Irv Bernstein's 12.8 antibody from the Fred Hutchinson Center, Berenson left the Hutch to found a cell separation company. Miller estimated they needed to raise $2,500,000 to open the doors. A decade earlier Berenson would have applied to the federal government for a grant to continue his research, or convinced one of the pharmaceutical giants to fund his work. But like the homely secretary who takes off her glasses and shows the world she's a beautiful woman, Genentech's success had made biotech sexy. The place to go for money was the world of venture capital. Miller wrote a succinct five-page proposal and began searching for initial funding

Genentech had turned on the spigot. Money for biotech start-ups was flowing from venture capital companies like a great waterfall. People standing at the bottom holding the right bucket could fill it quickly.

While venture capitalists didn't quite have the elan of riverboat gamblers, they were far more willing to take risks than traditional investors. Many people trace the beginning of the venture capital industry to 1909, when David Starr Jordan, the president of Stanford University, invested $500 in the development of electronic engineer Lee deForrest's audion tube, which supposedly would amplify an electronic signal. Three years later deForrest and two colleagues from the Federal Telegraph Company watched with excitement as a housefly thundered across a single sheet of paper. DeForrest's rudimentary vacuum tube had amplified the sound of the fly's footsteps 120 times, giving birth to the modern electronics era— as well as the world of venture capital.

Until the late 1960s most venture capitalists were individual investors, most often using their own money to stake a claim in a new company. In return for taking a big risk they expected a big payoff. But as it became too expensive and too risky for the lone banker to finance new businesses, they were replaced by firms who raised capital from outside investors and used those funds to support promising start-ups.

Venture capitalists were entrepreneurs with other people's money. But unlike individual investors, the venture capital firms actively participated in the management of the company. Like the early homesteaders who claimed land on the condition that they lived on it and developed it, VCs formulated business plans, hired executives, and served on boards of directors.

Many of the venture capital firms were located in the fertile farmland area just outside San Francisco once known as "The Valley of the Heart's Delight." Coincidentally, this was right across the San Francisco Bay from a small town in which I'd been raised, Pleasant Hill. While there was still some farming being done there, we knew it as a rapidly growing industrial area. I was living half a world away, in Belgium, when Berenson and Miller began searching for funding for their company.

Due to its proximity to several major universities, as well as the fact that electrical engineers Dave Packard and Bill Hewlett had founded Hewlett-Packard there in 1938, the Santa Clara Valley had become the center of the American electronics industry. In 1972 it was dubbed Silicon Valley by writer Don Hoeffler, the name deriving from the fact that the semiconductors that had made the great advances in electronics possible are made from the mineral silica. After Intel successfully produced the first microprocessor, the "computer on a chip" in 1971, the computer industry grew in the valley as easily as prunes and apricots once did. The venture capital firms originally came to the valley to support the new computer industry—but with the successful launch of Genentech they began growing another crop.

The difficulty in raising the initial financing was not the science—many venture capitalists lacked the knowledge even to understand what was being talked about—but rather the size of the market. If the system worked, how much money was it worth? How many bone marrow transplants were done each year? Would it be profitable? Venture capitalists weren't interested in creating companies that would save a few lives; there just wasn't enough profit in that. They wanted technology that would save a lot of lives. Big diseases meant a larger market. Miller and Berenson were competing for start-up dollars against potential treatments for far more prevalent diseases.

It turned out to be far more difficult to raise the start-up dollars than Miller had anticipated. In the past cell separation technology had been used almost exclusively for diagnostic purposes. Certain cancers, for example, were diagnosed by examining cells. Miller's business plan focused on cell therapy, using cells to cure disease. This was a relatively new, still unproven and controversial concept. Patients traditionally had been treated with drugs, not cells, and venture capitalists were hesitant to fund a new treatment protocol.

Among the venture capitalist firms Miller approached was Kleiner Perkins Caufield & Byers, who had funded his earlier ventures. Kleiner Perkins is one of the world's most successful venture capital firms. In addition to raising the original $100,000 needed to start Genentech, Kleiner Perkins has founded Compac, Sun Microsystems, Tandem, Netscape, even Sportsline.

Miller's proposal eventually reached a young associate in the life science area, Joe Lacob. In a typical year Lacob received a thousand new business proposals. Of that thousand he examined about fifty and, at most, supported investing in two or three of them. The fact that Rich Miller was involved with this project guaranteed it would receive Lacob's attention. Lacob respected Miller's judgment, which he described as a good combination of big-city street smarts tempered by clinical knowledge.

Lacob liked the concept immediately. The potential value of the cell separation market was estimated at about $100 million annually; if Berenson's Ceprate device really worked, it would work big. For almost six months Lacob and Brock Byers, who ran the life science area, debated whether or not to invest. They met with Miller several times. They were walking a very narrow ledge and, depending on which direction the wind blew from, they could have fallen on either side.

Lacob decided to jump on board and found the company himself. He felt he needed a new challenge and this appealed to him. Lacob understood the business of science. He had grown up in New Bedford, Massachusetts, the son of a stitcher at a leather goods plant who worked his way up to foreman. After graduating from the University of California with a degree in biochemistry, Joe Lacob had managed a small HMO center. Eventually he returned to Stanford and earned his MBA just as the biotech industry was exploding. He was hired by one of the real go-go start-ups, Cetus, where he was given responsibility for the marketing and commercial development of the drug that was supposed to cure cancer, Interleukin-2.

Lacob left Cetus to try to found his own biotech company. While attempting to raise start-up capital he met Brock Byers, who eventually offered him a job in the fledgling life science area at Kleiner Perkins. Lacob had been very successful in this position, but he missed being in the center of the action. He still wanted to build his own company. In Miller's business plan he recognized the opportunity to do that.

Every new company is a business experiment. It's a brew of people, place, and product, and you mix it all up and hope that you've found the right formula. You can play it safe and lessen the odds, but there are no guarantees of success. This new company was a particularly risky experiment. It was a new technology in a developing industry. It would have been easier, and probably financially prudent, for K-P to turn it down. But Joe Lacob's decision to run it, to serve as the company's first chief executive officer, made it happen.

Sandra Horning, Rich Miller's wife and herself a noted oncologist, named the company. CellPro. Cell professionals. Berenson decided that CellPro would be located in Seattle, Washington, where he had ready access to the people and technology at the Hutch.

While Berenson continued developing the technology, Lacob began building the company. Eventually CellPro would grow to 160 employees occupying its own building in Seattle's fabled east side Technology Corridor, but the company actually started with five people in what had been the morgue in an abandoned hospital owned by the Hutch. At most it was 1,200 square feet, consisting of the large laboratory and several very small offices. The only remnants of its years as the hospital morgue was a single stone slab where autopsies had once been performed. It was a dreary, isolated, cold place; the lab had no windows, making it impossible to tell day from night. CellPro was the only tenant in the building, so it was eerily quiet. Somehow the place seemed more fitting for the experiments of a mad scientist than for the initial work of a company planning to revolutionize medical treatment.

There had never been a biotech company quite like CellPro. It didn't fit any traditional model. The classic biotech start-ups like Genentech, Immunex, Amgen, Cetus, and Biogen developed drugs, not medical devices. As CellPro was among the very first biotech *device* companies, the development program was unusually complicated—CellPro was creating a medical device with a biological component to be used for therapeutic purposes. That required a broad range of highly specialized people working together—biochemists, cell biologists, and experienced lab technicians—as well as mechanical, electrical and software engineers to design and construct the devices. Cell biologists and engineers don't normally work closely together. Scientists and engineers don't speak the same language. Creating a good working relationship between them often proved to be a challenge.

CellPro wasn't recycling old equipment or modifying an existing system; there was nothing in existence like the Ceprate device. It had to be created from the imagination of CellPro's director of product development, Tony Goffe. Designing a medical device requires the skill of a master carpenter, the creativity of a small-town appliance repairman and the knowledge of a scientist. That basically describes Tony Goffe. He was a black man who spoke with a distinguished British accent. Tony was very likeable, always smiling, frequently laughing, even-tempered, quick-witted and energetic. He was also a great problem solver; when something had to be done he'd organize a group of scientists and engineers and they would just bang away at it until it was finished.

The first Ceprate platform looked like something that had been jury-rigged in the back of a garage out of baling wire and chewing gum, the work of a do-it-yourselfer gone mad. It consisted of a vertical slab from which hung several tubes that ran into a little plastic cylinder; this cylinder was the column in which avidin mated with biotin. Another tube ran out the bottom of the column into a waste bag. But in that original device were just about all of the elements that would be part of the final Ceprate system.

As Berenson worked on CellPro's first major project, a method of harvesting stem cells from bone marrow, Joe Lacob began searching for a permanent CEO. A search firm eventually identified two very strong candidates. The first choice of the board was Tim Anderson, the president of the Fenwal division at Baxter Healthcare. And Anderson appeared to be very interested in the job.

I knew Tim Anderson quite well. While I never reported directly to him, for a time before heading Baxter's marketing operations in Europe I had reported to the executive directly below him. I'd spent a considerable amount of time in large Baxter meetings with him. Tim Anderson and I could not have been more different—as much as I was fascinated by technology, he was passionate about the numbers. He could find the meaning hidden beneath mountains of numbers probably better than anyone I'd ever known. Obviously I had no idea that he was considering leaving

Baxter to become CEO of CellPro; at that point I had never even heard of CellPro.

Initially Lacob was concerned that Anderson would not want to live in Seattle, but Anderson assured him that relocating would not be an issue. Anderson was given CellPro's business plan, he visited the fledgling operation, and the technology was explained to him. He had access to all of CellPro's intellectual property. Lacob wanted to hire him, the entire board wanted him; no secrets were kept from him. Finally, Lacob made him a substantial offer.

Anderson turned down the job, explaining that his wife did not want him to accept it. Lacob wasn't terribly upset; he would much rather Anderson turn down the job than accept it and leave after a few months. This was business as usual. Anderson returned to Baxter and that might have been the end of it.

Might have. Except that approximately six months later Tim Anderson sublicensed Curt Civin's patents from Becton Dickinson, and Baxter went into the stem cell separation business.

Baxter Healthcare is a medical Goliath, manufacturing and marketing a range of products for blood and renal therapies, cardiovascular medicine, and intravenous solutions and disposables. Founded in 1931, Baxter is a *Fortune* 500 company that reported $6.1 billion in sales in 1997. They operate 175 facilities in 50 countries, have 42,000 employees and sell products in 112 nations.

Compared to that CellPro had just moved out of the morgue into its first office space. It had fewer than 30 employees and still hadn't made a single sale. But CellPro had all the virtues that a start-up brings to the marketplace, an entrepreneurial spirit, the ability to move quickly without fighting through layers of corporate bureaucracy, and a single product on which survival depended.

Tim Anderson and Baxter did absolutely nothing illegal. Anderson was never asked to sign a confidentiality agreement, and never did so—which in retrospect was a mistake. But he was given access to all of this information because CellPro very much wanted to get a CEO of his stature. Legally, Anderson was free to do anything he wanted to do with that information. In fairness, it is entirely possible that Baxter had fully intended to go into the cell separation business and had done its own research without anyone at CellPro being aware of it. Baxter might already have been in negotiations with Becton Dickinson for the patents.

It is possible. Baxter had already set up a new business unit, called the immunotherapy group, to develop products to be used in transplantation. So they were certainly very interested in cell separation technology—in fact they had developed a rudimentary magnetic cell separation device, the Mag-sep, to purge tumor cells from bone marrow. But it never worked very well.

So it is possible that this was simply a coincidence.

But it was quite a coincidence that Baxter acquired those patents after Anderson's visit to CellPro and then became our chief competitor. As Joe Lacob once said when discussing Anderson's role, business is business, but it has always seemed to be a slimy thing to do.

After Tim Anderson turned down the job, the board hired Chris Porter, a Ph.D. from the pharmaceutical giant Pfizer, where he'd been a vice-president and general manager of a major medical device subsidiary. Chris Porter was very well regarded. He wasn't a scientist, his background was scientific engineering, but he was an excellent choice for the company at that moment. Chris Porter successfully managed CellPro's early growth. In the critical first two years he directed the engineering and development of the Ceprate system. Eventually he realized that CellPro needed someone with medical device marketing experience to help develop marketing strategy and possibly form corporate partnerships with larger companies—Baxter, for instance. That's when I was contacted by Nathan Berry.

After the Baxter meetings concluded I flew to Seattle and met with Chris Porter, Ron Berenson, and several other executives. I was very impressed by Berenson and Tony Goffe—the guy who worried me was the chief financial officer, a rather quiet man named Larry Culver. While everyone was animated and excited about this technology, Larry sat quietly, looking very worried. I didn't feel as if I connected with him at all. I toured their cramped facilities. They showed me slides that demonstrated the technology. It was very impressive. I knew a bit about the market because Baxter had set up an immunotherapy group. As head of Fenwal, Baxter's European operations, part of this group reported to me. The technology was in an embryonic stage. About all I knew was that there was such a thing as stem cells and the theory was that you could pick them out by positive selection or depletion.

All of this happened very quickly. The thing that made me most suspicious was that it seemed too good to be true. It was difficult to believe that the perfect job for me in the perfect location would pop up completely unexpectedly at the perfect time. And the job did intrigue me. This was potentially spectacular technology. It was a means to treat patients on a cellular level. If it could be developed millions of lives would be changed. The opportunity to help make that happen was thrilling. I remember Berenson telling me, "It's almost as if you've spent your entire career preparing for this job."

The most difficult aspect of making career decisions is that it's impossible to know if you're on a track running in a big circle or on a path that leads somewhere. Success is so much a mix of hard work and opportunity and just simple good luck. Decisions have to be made on the basis of opportunity, compensation, lifestyle, and finally you just take a shot. It feels right. The one thing that no one ever considers is: Will this be the job that saves *my* life?

Eventually I met several board members. And finally they offered me the job. The salary was slightly more than I was earning at Baxter, but part of the compensation package was an excellent stock option plan. If CellPro was successful, that stock would be worth several million dollars.

I wanted to take the job, but the final decision would be made by the whole family. Perhaps because we moved so often we had become very close, closer than most of the families we'd known back home. I have a singular relationship with each of my boys, but I consider both of them my best friends. Jamie and Ben are very different people and have always been that way. Jamie has been independent since before he was born— Patty was in labor more than nineteen hours before he finally decided it was time. And then after he was born he just wouldn't pee. I had never known that that was an indication of potentially serious kidney problems in babies, but the doctors were very concerned about it. We were packing him up to send him to the Children's Hospital when he decided it was time. He's been independent his entire life.

Jamie is very much like I was when I was a kid, Ben is more like I am now. Ben just popped out, fat and bald as a billiard ball. Ben has always wanted input; in fact, he's always demanded it. Like younger brothers throughout history, Ben is the charmer.

We've established several family traditions, but the one we most enjoy is our Friday night movie. Friday night the four of us, and anyone else we can drag in, make popcorn and watch a movie. What makes this a little different is that we often watch the same movies many times, and as we watch we quote lines from these movies. It's our version of *The Rocky*

Horror Picture Show, and, in fact, lines from various movies have become our family shorthand. For example, when we were considering how to inform Baxter I was taking the CellPro job Patty suggested Danny DeVito's line from *Other People's Money:* "I'm doing my job. I'm a capitalist."

With the exception of an occasional visit to Patty's family, Jamie and Ben had never spent any time in the Northwest. But after spending a few days exploring the area they decided we could live there. Or, as the great *Fletch* said, "Well, there you have it." Meaning everything is acceptable.

Patty has always loved it there. She once told me that wherever she goes, she carries Mount Rainier with her in her head. The fact that Seattle was on the water made it even more enticing to both of us. We could put our sailboat back in the water. So, the Murdock family agreed to accept the job.

Baxter's management was not happy when I informed them that I'd accepted another job. They felt they had made a substantial investment in me and I was bailing out on them. For a time they threatened to refuse to move us back to the States, based on the premise that I was no longer a Baxter employee. But after I agreed to help formulate a transition plan and integrate my replacement we reached an uneasy truce.

One person I never heard from during this period was Tim Anderson. Of course, I did not know at the time was that he had been offered the job of CEO at CellPro and had turned it down. But we would soon meet again, this time under less than pleasant circumstances.

3 I WAS TO WAGE TWO TITANIC BATTLES AT CELLPRO, one for my own life and the other for the life of the company. And much like the elegant double helix of DNA, the two would be inexorably entwined.

To understand the fight for the survival of CellPro it is important to first understand how the government historically has wrestled with the concept that corporations might be able to own living organisms, a morally ambiguous idea with staggering implications. In 1972 a microbiologist working for Exxon had attempted to patent an oil-eating microorganism he had created. Theoretically these organisms could be spread over an oil spill and would safely devour it. But when he attempted to patent it, the U.S. Patent Office turned him down, "on the grounds that living things are not patentable subject matter . . ."

Exxon appealed to the Patent Court, which reversed the decision of the Patent Office. Eight years after the original patent was filed the case reached the Supreme Court. The question of whether companies could patent and own the elements of life as they did engine parts or yo-yos had never been addressed. Could a company own the rights to a chemical that kept the human heart beating? Was there a difference between an antibody that saved lives and the sole of a shoe? Can living matter be patented? There were important ethical as well as legal considerations. Billions of dollars were staked on the answers to those questions.

The United States Supreme Court answered the legal question in 1980. In *Diamond* v. *Chakrabarty* the court supposedly was deciding whether the "invention of a human-made genetically engineered bacterium capable of breaking down crude oil, a property which is possessed by no naturally occurring bacterium" was patentable; in fact they were deciding whether or not biotechnology would become an industry.

"(L)aws of nature, physical phenomena and abstract ideas are not patentable," the court decided, specifically pointing out that "a new mineral discovered in the earth or a new plant found in the wild is not a patentable subject matter." Newton could not have patented the law of gravity. Einstein could not have been given a patent to his explanation of all nature, e = mc². But, the court continued, "the patentee has produced a new bacterium with markedly different characteristics from any found in nature and one having the potential for significant utility. *His discovery is not nature's handiwork, but his own.*"

The patent office was ordered to issue this patent. Life was a commodity. The biotech gold rush had begun.

The ethical question remains to be settled.

The concept that scientists could create new life forms had been staggering to me. Centuries of science fiction writing had become fact. As a young businessman selling the end products of medical research I knew that eventually this technology was going to find its way into the marketplace. That was obvious to everyone. The vast amount of money necessary to support this research would have to come from private industry—and those companies that invested expected a return on their money.

Until that time university labs had always operated completely independent of industry, pursuing knowledge rather than profits. The federal government had made that possible by funding most of this research. It was costing taxpayers several billion dollars a year, and we weren't getting much back on that investment. That system just didn't make any sense. Many promising discoveries made by scientists working in academia never got out of the laboratory—or ended up on the shelf—because private industry would not risk millions of dollars to try to develop technology for which they couldn't license exclusive rights.

But in 1974 Stanford petitioned the National Institutes of Health for the right to patent Cohen and Boyer's recombinant DNA technology. With great trepidation, knowing the government might be accused of giving away public property to private industry, the NIH decided to grant tax-exempt universities the right to patent and license for-profit government-funded research. With that decision, as easily as waving a magic wand, university laboratories were turned into potential profit centers. A lot of people were concerned about the impact this would have on academic freedom; even NIH director Donald Frederickson wondered, "Are we turning university labs into industrial labs?"

At just about the same time, with little discussion or debate, Congress institutionalized the NIH decision by passing an amendment to the Patent and Trademark Act, one that was to become known as the Bayh-Dole Act, after its primary sponsors, Senators Birch Bayh and Bob Dole.

The objective of Bayh-Dole was to substantially increase the exploitation of discoveries made in government-sponsored projects by allowing universities to patent these inventions and license them to private industry. It wasn't a very sexy subject, and with the exception of lobbyists no one paid too much attention to this bill. It was just another piece of easily overlooked political business—except that it changed history. It made the fruits of some of the finest medical research laboratories in the world available on an exclusive basis to major corporations. It generated billions of dollars in licensing payments to these universities and led directly to the development of innumerable drugs that extended life and restored good health. It made possible so many of the products that transformed medical research into a business.

I might not even have been aware of Bayh-Dole when it passed. I certainly don't remember reacting to it. Nobody threw a big party to celebrate. And while through the next decade I was vaguely aware that it existed, I didn't know its many provisions—and I never suspected it would become the battlefield on which my company would make its last stand.

Everyone benefited from Bayh-Dole. The university would receive a royalty, private industry would not have to do the expensive primary research, new companies and jobs would be created to exploit these discoveries, and the public would benefit from the new products that would reach the marketplace.

Just as the right to collect tolls had once served as an incentive to build roads, the ability to patent discoveries turned biology into a business. The old academic warning, "publish or perish," was replaced by "patent or perish." A tidal wave of small start-ups financed by venture capital firms licensed patented technology from universities and began developing new products. Universities built a pipeline to private industry worth billions of dollars. Scientists became businessmen, often leaving the security of academia to found a company based on their work. In a sense, this was one of the greatest garage sales in history, as scientists pored through their old notebooks looking for anything that might interest investors.

Prior to Bayh-Dole, Curt Civin and Ron Berenson might have simply announced their discoveries and basked in glory—like Jonas Salk. But this was a different time. In February 1984, under the terms of Bayh-Dole, Johns Hopkins University filed a patent application for Civin's discovery that claimed the rights to just about any use imaginable of stem cells—even though Civin had failed to demonstrate the ability to use it—as well as any substantially pure collection of stem cells gathered using any technique. It was almost as if the British had claimed the entire continent of North America because the Pilgrims had first planted their flag at Plymouth Rock.

It was a very broad claim. Hopkins was asking for ownership of the stem cell, meaning that anytime, anywhere for the seventeen-year life of the patent if anyone wanted to use stem cells in their research or product, they would have to obtain permission from the university. But that patent application was not at all unusual or unique. That was very much business as usual. Other universities were applying for equally broad patents for discoveries made in their laboratories.

Unfortunately, the United States Patent Office wasn't prepared for this deluge of biotech patent applications. Maybe the world had changed, but not the system. Cutting-edge biotechnology was fed into the same system that issued patents for bed rollers. That wasn't the fault of the patent examiners, who had been thrown into a difficult situation. In some cases they simply lacked the scientific knowledge necessary to make proper determinations. Pretty much the same person who issued patents for clothespins was also working on biotech patents. And this field was so new the patent law hadn't been settled. Several cases were slowly moving through the court system on their way to the Supreme Court. The patent examiners were using the instructional manuals for biplanes to try to fly jets. It was often a rocky flight.

Obviously I'm biased, but I don't believe Curt Civin should ever have been granted such broad patents. Certainly Dr. Civin should have received a patent on the antibody he discovered and the hybridoma, the biological manufacturing plant that produced it, which he had made in his laboratory. As the court had said, this discovery was not nature's handiwork, but his own.

But rather than limiting his patent to his discovery, the patent office gave him the whole universe of stem cells. Basically, the way these patents were written, if someone used rocks and bottles to separate cells and ended up with a suspension containing 90 percent stem cells, they would be infringing Civin's patent—even though they had not used his or any antibody.

Hopkins licensed these patents to Becton Dickinson. BD eventually sublicensed the therapeutic applications to Baxter Healthcare. I've never known exactly when CellPro's original management became aware of the existence of these patents. What is certain is that management believed completely that these patents presented no serious legal problems.

Among the very first people Joe Lacob contacted when CellPro was founded was a highly respected figure in biotech history, Tom Kiley. Kiley, at one time a patent litigator, had represented Genentech from inception and served as its first general counsel. Having earned a fortune when Genentech went public, Kiley was sitting on several corporate boards and helping new companies get established. Lacob brought him to CellPro to negotiate the license for Bernstein's 12.8 antibody and Berenson's avidin-biotin technology from the Hutch.

The Hutch had never attempted to patent its antibody, and made no claims to any patent rights. All they offered was an exclusive license to use the 12.8 antibody. Eventually CellPro licensed those rights to the technology and the antibody for $50,000, plus an approximately 5 percent royalty. In addition to the license, though, what came out of the deal was Tom Kiley.

After looking at CellPro's prospects Kiley asked to join the board of directors. Lacob was thrilled to have him. Kiley carried with him the mystique of Genentech. His presence at CellPro immediately gave the company credibility. It certainly helped attract the attention of the Wall Street bankers whom Lacob would eventually need to court.

Lacob eventually asked Kiley to take a look at the Civin patent. Kiley was firm in his opinion that Civin's primary patent was invalid because Civin had published his discovery before receiving his patent. CellPro was using its own antibody in the technology Berenson had developed. In fact, CellPro had the opportunity to sublicense the patents from Becton Dickinson, but BD demanded more than a million dollars for the therapeutic rights—the rights to use Civin's antibody to harvest a purified collection of stem cells to treat patients—much more than any start-up could afford. Particularly to buy rights nobody thought were needed. Given the level of concern about those rights, it would have been like using chemotherapy to cure a headache.

Kiley didn't equivocate at all. Not in the slightest. He was a respected patent attorney, a Genentech man, and there was absolutely no reason to doubt him. Later, Kiley hired a friend, another highly respected patent attorney named Coe Bloomberg, from the preeminent intellectual property firm Lyon & Lyon to render an outside opinion. This was an area in which Lyon & Lyon had considerable experience: in 1986 they had won the first federal biotech patent infringement lawsuit, a case that focused on monoclonal antibodies. Bloomberg agreed completely with Kiley. After reviewing the history of the Civin patents he concluded that the key patent claims were invalid.

The patent issue did cause some concern when Lacob began looking for additional investors. At that time Kleiner Perkins preferred to syndicate its deals, meaning they shared the financial risk with other venture capital firms. It cost slightly more than $2 million to found CellPro, of which K-P put up about $800,000. Within a few years $2 million would seem like nothing, little more than lawyers' fees (actually a little less), but Lacob wanted less financial exposure. Several other venture capital firms "kicked the tires hard," according to Lacob, "but walked away because of the patent issue." Eventually three other venture capital firms agreed to participate in the initial financing. In April 1989, CellPro officially went into the business of cell separation.

CellPro's burn rate, its fixed expenses, increased every month as the company prepared to begin the initial clinical trials of the Ceprate device. To meet those expenses Lacob returned to the venture capital marketplace in the spring of 1990 and raised an additional $8 million in secondary financing, a "series B" as it is called. It became apparent during this second round of financing that potential investors remained nervous about the patent situation, although Coe Bloomberg's strongly worded opinion had allayed most fears. This was the kind of nettlesome problem that often popped up and was usually settled rather easily.

No one at CellPro was happy that Baxter had licensed these patents, but personal feelings had to be put aside. Business is business. It was obvious that the best way to handle this problem was to forge some sort of mutually beneficial deal with Baxter. And there were many ways in which that might be done. While CellPro was years ahead of Baxter in the development of a stem cell separation system, Baxter had an extraordinary overseas marketing and distribution network. It made great business sense for both companies for Baxter to become the distributor of the Ceprate system outside the United States. Baxter could save a fortune in research and development costs, and CellPro would save a comparable fortune by not having to set up its own worldwide marketing operation. A corporate partnership would serve the interests of both companies.

Clinical trials leading toward FDA approval of the Ceprate system began in June 1991. This was the first major step on the journey to the marketplace. A month later, on July 25, 1991, Baxter executives Russ Hayes—who reported directly to Tim Anderson—and Bill Lake visited CellPro to initiate discussions about forming some sort of partnership. The meeting went very well. Apparently Hayes told Chris Porter that Baxter had been watching CellPro for quite a while and was extremely impressed with its progress. Baxter believed that CellPro's technology was complementary to its own, as opposed to competitive. In fact, Hayes explained that Baxter was not wedded to a specific technology, and that he felt products would be based on technologies that had not yet been developed, or on combinations of technologies.

Chris Porter reported to the board that the meeting had been non-confrontational and there had been no mention of the Civin patents. At the end of this meeting Hayes and Lake agreed to sign confidentiality agreements in anticipation of moving forward with negotiations.

Those negotiations proceeded very slowly. About a month later Chris Porter spoke again with Hayes and Lake. While the conversation was pleasant and seemed promising, in fact it identified an issue that would never be resolved. "Hayes wants to outline the framework of an agreement . . ." Porter wrote in his note to the board. "He was also talking about the possibility of an equity investment. They do want global rights to the product.

"I said that the U.S. was going to be a problem because we want to sell the device directly."

The U.S. market was the place CellPro would make the largest profit on the Ceprate system. It was the crown jewel, the one thing CellPro could not afford to give away in negotiations and remain a viable company. Baxter knew the value of the U.S. market as well as CellPro did, and wanted it. What no one at CellPro understood at that point was just how badly Baxter wanted it.

I was thrown headlong into the fight a lot more abruptly than I had anticipated. I'd been at CellPro only three months, barely long enough to find the men's room, when I was promoted to president. In only twenty-nine months, CellPro had been successfully transformed from Berenson's technology and Miller's five-page proposal into a public company with an estimated market value of $100 million. Wall Street and the financial media celebrated CellPro as a model start-up. It was a major success story. But just below the surface the waters were roiling. I had stepped unknowingly into a serious management problem.

Chris Porter had done an admirable job building the company during his tenure as CEO. He had turned raw material into a marketable product. To outside observers CellPro appeared to be a vibrant, well-managed company poised to become a biotech industry leader. But internally the company was being ripped apart. Key members of management differed greatly over the future direction of the company. There was a real power struggle going on between the scientists and the engineers. At one of the first meetings I attended a senior scientist accused a senior development executive of hiding or falsifying data, angrily threw down his papers on the conference table, and stormed out of the room. *Wow,* I thought, *what have I gotten myself into? What's going on here?*

Chris had not been able to resolve that dispute. He tried everything to bring the two sides together; he'd even brought in a psychologist, but he was unable to effect a compromise. Eventually the board of directors had to make a choice. As the future of the company depended on the successful development of the biological product, it supported the scientists. But in the fight Chris Porter had lost the confidence of several key board members. The board decided that he had to go.

A few days before Christmas Joe Lacob told Chris that they wanted to make a change at the company. "The timing was awful. This was like four days before Christmas," Lacob remembers. "I'd read about situations like that, and I'd thought, 'God, I'd never do that,' but I had no choice."

The board hired a search firm to find an executive with biotech experience. In the interim, they asked me to run the company. I was flattered and confident, but it was a bittersweet situation. I liked Chris Porter very much. I would have been very happy working for him. Chris and his wife had welcomed my family when we arrived in Seattle. We'd bought a house only a few blocks away from his family. We had been guests at his home for Thanksgiving.

I didn't want Chris to think that I had undermined him. So after these changes were announced I went into his office, looked him right in the eye, and said, "I'm really sorry about all this, Chris, but I want you to know that I did not do this to you."

"I know that," he responded, "I understand the situation."

On December 19, 1991, it was publicly announced that I had been appointed president of the company. I was forty-four years old. To reduce the impact on our stock, officially Chris was promoted to vice chairman and chief executive officer, but I began running the company. He handled a difficult situation with class and grace. We were in the process of building our manufacturing facility and he agreed to head that project. And did a magnificent job.

As CellPro's new president there were many things I needed to get done, but among the most important was negotiating a deal with Baxter. I suffered no illusions that it would be easy. On my bookshelf sat several awards I had received from Baxter for achievements in technical development, sales, and marketing. I had my Baxter gold watch. I had worked with several of the people with whom I'd be negotiating. So I knew how tough it would be, but I firmly believed that we ultimately would be able to reach an equitable agreement.

As it turned out, I knew nothing at all.

Our objective in these negotiations was to obtain a clean license to their technology without giving away our domestic marketing and distribution rights. If we were able to reach an agreement in which we gained access to their worldwide marketing network it would be an excellent deal for both companies.

We had finally decided to license the Civin patents, even though our attorneys had assured us that legally we didn't need them, because we didn't want to fight Goliath. I had been a cog in the Baxter machine and I had negotiated with them while selling HemaScience, so I knew enough not to want to get involved in a legal battle with them. Baxter had the multibillion-dollar bottom line to grind down an opponent our size. They could outspend us, outlast us, and outlawyer us. At times it seemed to me that Baxter had more lawyers than CellPro had employees.

Lawsuits did not frighten Baxter. I knew of an ex-Baxter employee who had left to found his own small company, only to be sued by Baxter and put out of business. He had claimed ruefully that Baxter was the only company he knew that treated its legal department as a profit center.

I don't think that was meant to be a joke.

While the legal battle we waged against Baxter might be seen only as a fight between two companies protecting their bottom lines, it was far more significant than that. We were scratching the legal surface of the future. The laws that so magnificently enabled us to grow and prosper as a society in the past have had to be stretched grotesquely out of shape to be applied to the new technologies. This was a confrontation between business and medicine, between stockholders and patients. It is a battle that is going to be fought often and bitterly as companies continue to discover means of preserving and extending life. This was about the value of life.

In late November 1991 I flew down to Irvine, California, to meet with Russ Hayes, the president of Baxter's immunotherapy group. I knew Russ from my days at Baxter, and had long been underwhelmed by him. He was very typical middle management, uninspired and plodding. I considered him a corporate hit man, the kind of manager who would happily go with the flow until his next assignment. It was my belief that Russ Hayes was simply a messenger, that the real decisions were being made by Tim Anderson, by then the president of Baxter's biotech group. I knew from experience that no significant decisions were made in this area unless they were blessed by Tim Anderson.

I was quite apprehensive about this meeting. I knew that a lot of people at Baxter were unhappy when I left. They felt I'd bailed out to join the competition. But I'd strongly made the case that this was not a competitive situation. Baxter was nowhere near having a product. I didn't expect to actually accomplish very much at this meeting, but I wanted to take Baxter's corporate temperature. I wanted to know how they felt about CellPro—and about me.

Coincidentally, Baxter's immunotherapy group was located in Irvine because that's where HemaScience had been. After acquiring us they'd basically moved into the neighborhood. So there were a lot of mixed memories for me as I walked into that building.

I wasn't so much nervous as I was curious. It hadn't taken me long to shed the corporate straitjacket I'd worn at Baxter. I was very much back in the start-up mode. Although I'd left Baxter only a few months earlier, the place felt so foreign to me that it seemed as if it had been years. But I had arrived with a lot of confidence. I didn't believe we *needed* to make a deal, but we definitely wanted to make a deal. If we failed, we still were several years ahead of them in developing a stem cell separation device and there was little they could do to change that.

Russ Hayes and I met alone in his office.

To my surprise Russ was very straightforward and open. Almost immediately he acknowledged that CellPro was indeed at least two years ahead of Baxter in the development of automated cell separation technology and told me that they wanted to make a deal. "I don't think there's any reason for two companies to spend so much research and development money in this area," he said, "so we have to figure out a way to put this together."

I was thrilled to hear that. Absolutely thrilled. But I knew it couldn't possibly be that simple. "Well, we'd like to do a deal with you," I said. "We'd be very interested in talking to you about Europe." Meaning we would be willing to give them the rights to distribute the Ceprate system in Europe—and probably the rest of the world outside the United States. But the United States was sacrosanct—we could not give away distribution rights to this country and remain a viable corporation.

The meeting had been very cordial—until that moment. "No," Russ responded, "we only do global deals now. We're not interested in regional deals." Basically he was telling me that Baxter intended to control this technology whatever they had to do.

"Boy, Russ, that's a real problem for us," I explained, "because you have to understand the crown jewel of the company is our U.S. distribution rights."

He agreed that this was indeed a problem—and then warned me how large a problem it could be. "I hate to be rattling the saber, but you know, the Civin patents are out there. God knows we don't need another lawsuit, but they're there. And that's something we're going to have to deal with at some point." He added that this would be a nonissue if we reached an agreement.

Perhaps a year later, after lawsuits had been filed, Russ Hayes was deposed, and he denied that he had mentioned the patents during this meeting. That just wasn't true. As I had written in my contemporaneous report of this meeting to CellPro's board, Russ had told me, or perhaps subtly threatened me, "they did not want another lawsuit, but that this was an important issue for them."

And as I also wrote in this report, somewhat ominously, "We will have to be very creative to solve this one to our satisfaction."

To my great surprise, in January we received a letter from Baxter offering us a nonexclusive license to the patents. This was an amazing offer. It was an opportunity to license the patents without giving away any rights. They asked for an up-front payment of $750,000 and a royalty of 16 percent of the price of the antibody portion of our disposable kit.

They didn't define precisely the terms of the 16 percent royalty payment, but we didn't really care. We thought they were asking for loose change to settle this big thorn in our side. I immediately convened a board meeting and we put together a counterproposal.

I sometimes wonder what would have happened if we had simply accepted that offer on those terms. Everything I know about Baxter makes me believe it couldn't possibly have been that simple. I feel certain that they would have done precisely what they'd done in the past—hold that offer in front of us like a carrot while leading us into the maze of negotiations.

But we will never know because we didn't accept it. We felt that we should receive some credit for the fact that we had enabled cell therapy. We'd done the work. We'd licensed our own antibody, developed our own automated device, and treated patients. We were out there helping to save lives. Baxter's total contribution to the value of these patents was nothing. They hadn't done a thing to make them worth a penny more than when they'd licensed them from Becton Dickinson. Their My-10 antibody wouldn't work in our system. There was nothing they could provide us that had any value to CellPro. They were doing nothing more than becoming a middle corporation, and we believed the value of our work should be part of the equation. All we were buying from them was protection from a potential lawsuit.

But it was still a lot less expensive than a legal battle. Our board did a cost analysis and responded by offering to pay a royalty "not to exceed 30 percent of the total value of the kit," and an up-front payment of $500,000 that would be applicable against royalties. This was simply our response to their letter, but we knew we would roll over in a second if they stuck to their original demands. It was a fair counteroffer. The royalty we proposed was above the prevailing industry average.

This is known as negotiating. In business, when an offer is made, you respond with a counteroffer and eventually an agreement is reached. Rarely is a first offer the final offer. But even as we sent them our response I couldn't help but feel anxious. I remembered my previous experience negotiating with Baxter. At HemaScience, there had been times when I'd felt certain we had a deal—and they would find cause to back away then change the terms. This offer just seemed too good to be true.

Baxter acknowledged receipt of our letter—and we never heard from them again. Our board was getting very antsy. Four months later, four months, I called Russ Hayes. He was as friendly as ever. "We're ready to talk business," he said, and we arranged another meeting.

I was expecting the same kind of meeting we'd had the first time, just Russ and I alone in his office talking it out. Two guys having a friendly discussion. But it was nothing like that at all. It was a very formal meeting. We were joined by a Baxter attorney and an executive from the business development side whom I'd known in Europe. The presence of the lawyer changed the dynamics of the meeting entirely. His job at that meeting was to serve as a not-so-subtle threat. It was obvious this meeting had been well planned, and it was not going to be pleasant.

Everyone sat rigidly in comfortable chairs that had been drawn into a rectangle. The other two men said very little. "Here's what we want to do," Russ told me. "We want distribution rights to your product outside the United States. And we're willing to accept nonexclusive rights to the United States." He specifically wanted exclusive distribution rights to Europe and Japan in addition to nonexclusive rights in America.

I was just about knocked out of my chair. This was exactly what I had long feared. Baxter playing business hardball. As Hayes continued I could feel the old HemaScience twinge. Man, I thought, these guys are trying to do it to me again. I couldn't help but think of a similar meeting I'd had with Baxter executives during the negotiations to sell Hema-Science. Smiling broadly, they had told me the terms of the deal to which we'd agreed had changed and then practically dictated the terms.

I wasn't about to let that happen again. Hayes was proposing that we would make the product and they would sell it. There was absolutely no way we could agree to a deal like that. The real profit for a company is made in marketing, selling the product to the end-user. The profit margin in manufacturing is minimal. The fact that Baxter agreed to accept nonexclusive rights in the United States had absolutely no value. In fact it was worse. They would have become our competitor—using our technology. More important, we could not have done an exclusive deal with another company if we already had a nonexclusive deal with Baxter.

In this situation nonexclusive had the effect of becoming exclusive. I took a deep breath and said, "Russ, you guys sent us a letter offering us a license and we sent you a counterproposal. I thought I was coming down here to talk about that."

Russ frowned. "We don't want to do that anymore," he responded. "The only way we're going to give you guys a license is under these terms."

I had gone to this meeting to find out Baxter's intentions. Well, I had the answer to that question. It was apparent to me that they intended to use the Civin patents to gain control of our technology (without paying for it) and then own the market. "Look," I said, "there're probably some ways we can work together here...." I didn't want to get into discussions about U.S. marketing rights. That was a total non-starter. As far as the meeting was concerned, it was game over. There was nothing more to discuss.

I wanted to get up and walk out, and maybe on some level I wanted to tell them off, but I didn't want to sever negotiations. I still believed there had to be some way we could make a deal. Instead, I went with them to lunch. I remember sitting in a restaurant, listening to Russ Hayes talk excitedly about what a great partnership this was going to be, feeling as if I had just gotten a dagger in my heart.

Even if I had been tempted to agree to these terms, the fact that we were a public company made it impossible. If I had agreed to those terms, I'm convinced that our stockholders would have sued me for mismanagement, and they would have been right. I wanted to give Baxter Europe, and if they had pressed I would have given them the rest of the world outside the United States—but they wanted the entire universe.

At that meeting I'd told them, "You need to put your position in writing to me so I can take it to my board of directors."

The next day a letter written by the attorney, under Russ Hayes's signature, was faxed to my office. I leaned back comfortably in my chair and started reading it. I immediately sat up. When I finished reading it I read it again. I couldn't believe that Russ Hayes had actually permitted this letter to be sent under his authority. "Baxter is not interested in pursuing CellPro's counterproposal," this letter read, "nor is Baxter interested in pursuing a simple licensing of its CD-34 technology [the Civin antibody] for an up-front payment and running royalty....

"We propose the following arrangement whereby Baxter would license CellPro under the Civin and related patents....

1) CellPro would grant Baxter exclusive distributor rights for its products in Europe and Japan.

2) CellPro would grant Baxter nonexclusive distribution rights in the North American market, the United States, Canada, and Mexico.

3) CellPro would pay Baxter a running royalty on products not sold by Baxter equal to the royalty Baxter is obligated to pay Becton Dickinson under its license. . . ."

All this for rights we did not believe we needed. The original offer was dead. Hayes was telling us that the only deal in which Baxter was interested required us to give them worldwide distribution rights—in return for patent rights applicable only in the United States. Because Civin had published the results of his work, Hopkins was prohibited from filing for international patents. Whatever patent rights Baxter held stopped at the American borders, so this seemed to me to be a clear violation of the anti-trust laws. An American patent is valid only where the laws of the United States are in force.

A patent issued by the U.S. Patent Office has absolutely no standing in foreign countries. And it is a violation of the anti-trust laws to use the powers granted by a U.S. patent as a sledgehammer to gain rights in foreign countries. This letter stated clearly that Baxter intended to do just that. Playing hardball in negotiations is one thing, but I considered this to be a form of intellectual property extortion.

This letter was a smoking gun. We had caught Baxter with their lawyers in our cookie jar. Although the attorney had written the letter, Russ Hayes was responsible for it. Essentially, his career at Baxter was finished.

The next time we heard from them John Osth, whom I'd known in Europe, wrote, "I will be the point of contact, replacing Russell Hayes, who has recently left the company." There was no further explanation of Hayes's fate. But this letter was an obvious attempt to purge Hayes's letter in case we went to court. "We sought," Osth continued, "but certainly did not insist upon a nonexclusive right to distribute CellPro products in North America, recognizing that CellPro would be free to appoint other distributors. . . . All of this was, in any event, negotiable." Osth claimed that Hayes's letter didn't actually mean precisely what it said, but the damage had been done.

Tom Kiley had called me as soon as he saw Hayes's letter. He was even more upset than I was. "Listen," he told me, "I'm convinced that one way or another we're going to end up in court with these guys. They've put us in an impossible situation. If we don't agree to their terms they're going to sue us, and if we do agree the company is destroyed. We don't have any options. So I'd rather go to court on our terms than theirs. I think we ought to hit them with a preemptive strike."

My worst nightmare was coming true. The last thing I wanted to do was get involved in a lawsuit with Baxter. Knowing that company as well as I did and having seen that corporate ego, the thought of suing them scared the hell out of me. "Well, Tom," I said, "we ought to convene the board and talk about that."

Kiley contacted Coe Bloomberg at Lyon & Lyon and asked him to prepare a lawsuit. By suing them first in Seattle, we hoped to gain the home court advantage. Lyon & Lyon's legal papers claimed that Baxter's patents were invalid, that even if they were valid we were not infringing on them, that Baxter had illegally misused these patents and had committed several anti-trust violations. But Kiley agreed not to initiate court action without the permission of the board.

Two other companies eventually did accept Baxter's terms for a license. Both Applied Immune Sciences and SyStemix paid $750,000 and an 8 percent effective royalty. The difference between these companies and CellPro was that we had invested millions of dollars and developed a marketable device. Neither of those companies was even close to a product. In fact neither company *ever* took a product to market. Their cell selection technologies were very different from ours and would be very difficult to make work clinically. They were never going to be viable competitors in this field and Baxter had to know that. They just took their money.

The board was divided about suing Baxter. Most of us believed that unless we could reach an agreement, it was inevitable that sooner or later Baxter would sue us, most likely later, after we had an approved product on the market and were making a profit. It made no sense for them to sue us until then. I really did not want to initiate a legal battle with Baxter. I'd seen them in legal action, and they played to the corporate death. What had surprised me was the level of animosity and how virulent they were in prosecuting a case. So I dreaded taking this action. But at that point I didn't see any way we could reach an agreement. The one thing they demanded was the only thing we couldn't afford to give up.

Tom Kiley took the hardest line. Kiley believed Baxter was threatening our survival. And Kiley's opinion carried great weight with me and several members of the board. Kiley was quite a character. He is not a big man in size but certainly is in personality. Kiley was the kind of lawyer who could argue both sides of a case and convince a jury that both sides were right.

Kiley loves to win. From fishing to clothes, he is relentless in his pursuit of excellence. I remember that not too long after having heart surgery he made a 140-foot bungee jump off a bridge. He certainly wasn't shy boasting about it—even Kiley admits that at times he can seem arrogant. Several months later Josh Green, our corporate attorney, bungee-jumped 150 feet off a bridge in New Zealand. And did so with great fear and trepidation, he admits. Then he sent Kiley a note playfully pointing out, "I guess I did a jump 10 feet more than you." Not long after that Kiley was at Victoria Falls in Zimbabwe. Not to be outdone, he made another bungee jump—of about 366 feet. He immediately sent Green a fax chiding him, "I just beat you by 200 feet, sucker!"

That's perfect Kiley. But he was so sure that Baxter was going to sue us that he insisted our lawyers check the courts in Delaware and Washington, D.C., to make sure they hadn't already filed the papers. While I often agreed with Kiley, this time I thought he was being paranoid.* I knew the Baxter culture very well, Baxter moved with small, steady steps; suing us would have required a great leap.

As strongly as Kiley felt about suing them, Josh Green was just as adamant that suing them was a terrible mistake. I liked and respected Josh Green. While Kiley had served as CellPro's legal advisor and done a lot of the initial negotiating, Josh had done the actual legal work necessary to start the company. He and Joe Lacob were best friends, and Lacob had given him a lot of responsibility. In addition to incorporating CellPro he had done all the legal work to take the company public. Josh lacked the bravado of Kiley, but he was an excellent attorney and even in the roughest waters steered with a steady hand.

Admittedly, there were times that I got pretty upset about his attitude. He strongly urged us to find a way to settle this dispute, so much so that at times I wondered if he was afraid of Baxter. But one day as we were flying back to Seattle from a meeting he told me the real reason he desperately wanted to avoid a legal confrontation. When Josh was growing up his father owned a small wholesale jewelry business, Leo Green &

---

* In support of Tom Kiley's concern about being sued by Baxter, the Federal District Court in Seattle would later find that CellPro had a "reasonable apprehension" of an imminent suit by Baxter.

Company. In the 1960s he had created a Christmas catalogue for his retail clients that allowed them to stamp their own names on the back and distribute it to their customers. A much larger manufacturer circulated a smaller Christmas catalogue to the same retailers—but allowed them to sell the products in that catalogue only if they did not do business with Leo Green. It was a clear violation of the anti-trust laws and Josh's father had sued them.

The lawsuit went on for almost eight years. It never reached the courtroom. Just before the trial was to begin, the larger company agreed to settle with Leo Green for cash and merchandise. Most of the cash went to the lawyers, and the merchandise wasn't even marketable. The lawsuit sapped eight years of energy from Josh's father and destroyed his business.

And, as Josh pointed out, his father had won. So Josh knew very well that the real effects of waging a legal battle were much more complicated than simply winning or losing. Josh didn't believe we were prepared to take such a drastic action against a multinational corporation with endless resources. He strongly advocated doing whatever it took to reach an agreement.

I think his exact words were, "Are we fucking insane? This is crazy. We're not ready to file a lawsuit." When there remains even a small chance for peace, he insisted, you don't go to war. Over and over he warned that at best litigation is a 50-50 proposition, telling us, "Listen up, guys, you can't guess what the housewife on the jury is going to decide. This is just an impasse, there's still a chance we can work out a deal here."

Kiley and Josh Green really battled over this decision. Kiley was a legal warrior. He believed that the best defense was a good offense. He just wanted to go out and blast them. After listening to both sides I decided that if Kiley believed so passionately that we had to jump these guys, I had to support him. It was not a particularly difficult decision; Kiley was one of the leading patent attorneys in the country, he'd written the brief on which the Supreme Court had based its decision in the *Chakrabarty* case, he was Genentech's first counsel. He had been present at the beginning of the biotech revolution, even spending nights sleeping on Swanson's couch. If Kiley was wrong about the Civin patents we were out of business anyway; we would just be delaying the inevitable.

Kiley and I wanted to file the lawsuit. Green, Ron Berenson, and our chief financial officer, Larry Culver, were against it. The rest of the board was neutral. However, simply by the force of his personality, Kiley convinced a majority that we had no choice—it was either sue Baxter now or be sued by them later. We decided to file the first shot.

On April 28, 1992, we filed suit against Baxter and Becton Dickinson in federal district court in Seattle claiming the Civin patents were invalid and had not been infringed upon by our technology, and asking for relief under the antitrust laws. We threw down the gauntlet. There was no turning back. We specifically did not include Johns Hopkins because, ironically, at that time doctors at the university were using the Ceprate device in several clinical trials. Hopkins was one of our best clinical sites.

I was terrified the day our legal papers were hand-delivered to the courthouse. Terrified? I was shaking. I'd supported Kiley but I knew Baxter. It's a bad analogy, but I couldn't help comparing this to the Japanese sneak attack on Pearl Harbor—and I remembered the consequences of that act.

We caught Baxter completely by surprise. I think they were shocked that we fired the first shot. Within days Tim Anderson contacted Joe Lacob and told him, "Hey, you know, this lawsuit, it's no good. We can solve this thing." The real problem, Anderson explained, was me. "Murdock came down and scammed us. He misled us."

I never understood exactly what he meant by that. Perhaps because I hadn't stormed out of that second meeting with Russ Hayes, I hadn't simply rejected his proposal or told him it was a violation of the anti-trust laws, he had misread our position. I have absolutely no way of knowing what happened, but I suspect Hayes had reported to Anderson that we were going to fold, we were on the ropes.

But instead of a white flag of surrender, Baxter received notice that we had filed our lawsuit. Somebody had to be the fall guy; it had to be somebody's fault we'd reached this impasse. Hayes was gone so Baxter decided it would be me. Business no longer was just business; it had become personal.

At that point the board agreed that Kiley would take control of the negotiations. Kiley and I agreed to meet with Anderson and a Baxter attorney at their corporate headquarters in Deerfield, Illinois. As far as I was concerned, we were flying into the heart of darkness. If there was any chance of reaching an agreement, we would know it after this meeting. The posturing was done, it was time to make a deal or go to war. In the five years I'd worked for Baxter I'd made this same trip many times. I knew my way around their corporate headquarters.

We met in an impressive paneled office in the executive wing. It was decorated in corporate power. Absolutely every object in that room had been selected with great care. Every sound was muffled by the thick carpet. It was an office designed to be intimidating.

Kiley was not the slightest bit intimidated. He always saw himself as a master negotiator and at this meeting he was very good. The problem was that he was used to dealing with people who came to the negotiating table in good faith. This was not the case here.

We really didn't know what to expect. "Thank you both for coming," Anderson began. And then once again he told us that we could license the patents. The price was an upfront payment and effective 8 percent royalty rate.

Kiley sighed professionally. "Look, we're here to negotiate that rate. We don't mind paying a royalty, but we would like some recognition that we've done all the work to develop this market."

Anderson had the executive ability to make it seem as if he was agreeing with us while turning us down. He made it perfectly clear that he was not going to negotiate terms—he was telling us what they were. That's the deal, he said basically, without ever saying those words, take it or leave it.

That was precisely the wrong way to negotiate with Tom Kiley. Kiley stood up and said, "Well, gentlemen, I guess the good news is that we're not going to miss our flight back." Then he started to leave.

I don't know who was more surprised, me or Tim Anderson. I got up to follow him, but Anderson seemed dumbfounded. I don't think he knew how to respond. "But, guys," he finally said, "this is only a few hundred dollars more per kit and it's something you can easily pass on."

This was edge-of-the-cliff negotiating. I don't think either one of them had the personality to step back. It was obvious that their positions, as well as their egos, had solidified. Kiley responded, "I don't think there's anything else we can do here unless you're ready to negotiate. And if we're not going to negotiate there's no reason for us to stay here."

Maybe I was the president of the company, but I was just there for the lesson. Kiley walked out and I followed him. The meeting had lasted less than an hour. But as soon as we were outside Kiley paused and lit a cigarette. For the first time I understood that even though Kiley seemed so smooth, so suave, so in control while in that meeting, it had been an emotionally draining experience for him. We went to the airport and flew home.

The war was really just beginning.

Several years later, our Vice-President of Operations and Chief Technical Officer, Dr. Joseph Tarnowski, was being interviewed for a new job by a board member of another biotech company. Coincidentally, the interviewer was very friendly with a member of Baxter's board and knew the entire story very well. Baxter's strategy in negotiations with CellPro was clear, this man had been told, and repeated it to Joe: "If we lose in court we'll work out a business deal. But if we win, if it's found that CellPro infringed our patents, we'll pound them into the dust."

"Pound them into the dust." That's what they wanted to do, and that's what they did. Admittedly, I made mistakes during the negotiations with Baxter. While their terms were onerous, there probably were times when we might have reached an agreement. We would not have been the same company and we would have been exposing ourselves to stockholder lawsuits, but it certainly would have been better to live with a bad deal than close the doors.

Unfortunately, I'd brought with me to Chicago all my previous experiences with Baxter. I knew Baxter too well. My firm belief, based on what I knew about that company, was that they would not have agreed to a deal on the terms outlined by Anderson. Every time I'd negotiated with them in the past, their strategy had been to lead me way down the road, then change the terms. I just didn't believe that they would have given us a license for an up-front payment and a royalty. I can't prove it because we turned down their offer, but I never believed it.

▷ Although officially I was running the company only until an experienced CEO was hired, I didn't hesitate to exercise control. The negotiation with Baxter was only one of several serious problems I had to deal with immediately. I also had to find a way to resolve the conflict between the scientists and the engineers that had led to Chris Porter's ouster.

This was the first test of my leadership ability. I was aware that everyone was watching me closely. I had the advantage of knowing that a compromise was not possible. I realized I had to pick one side—either the scientists or the engineers—and back that side completely. It was obvious that for the survival of the company the most important person was Ron Berenson. CellPro was going to succeed or fail based on our success in the laboratory and the clinic. So Berenson was the horse I decided to ride. I took Tony Goffe, who had done a fine job as head of development, to lunch and told him that while he had contributed a great deal, he had to leave. I was shocked at his response. "I understand," he said. "I knew it was coming and I don't hold it against you." He left as a gentleman, with a smile on his face and a valuable chunk of stock. And I had solidified my position as someone willing to make difficult decisions.

An important part of my job was making long-term strategic decisions. The company had been founded to exploit Berenson's avidin/biotin stem cell selection system, but the elegance of the Ceprate device was that it could so easily be adapted to harvest almost any other useful cell. Whatever cell was desired, we could select for it. There were numerous possible life-saving applications for this technology and every one of them was potentially a business for us. I had to decide where to apply our limited research and development resources.

We were involved in three major projects: we were making and selling laboratory columns capable of selecting a limited number of blood cells for use in research programs; we were in Phase III clinical trials with the Ceprate system for stem cell selection; and we were developing a method of harvesting fetal cells from maternal blood that could be used in place of amniocentesis in fetal diagnostic testing. The question was, Where did we go from there?

Good scientists, and we had a laboratory full of them, like to follow their curiosity. If I allowed that to happen we'd end up with groups of researchers spreading their time working on a variety of projects, and making little progress on any of them. We didn't have the resources for that. We had to pick specific projects and focus on them. Other applications, no matter how promising, would simply have to wait. I relied on our clinical investigators, the doctors in the field treating patients, to tell us what they needed.

Two hundred years ago, in London, Edward Jenner learned from a milkmaid that people who had had a specific strain of cowpox did not get smallpox. He theorized that by inoculating people with cowpox, he could make them immune to potentially fatal smallpox. His experiments proved his theory was correct. The path from observation to experimentation to the creation and distribution of a smallpox vaccine was short and direct. Medical research no longer functions that way. The cost of pursuing every dream is prohibitive. So it's become a business of demand and supply. There is a certain symmetry to the numbers: Those medical problems plaguing the most people can generate the greatest financial return, so they receive the largest percentage of research dollars. As a result the largest number of people receive the treatment they need.

We were hearing a strong demand from the field for tumor cell purging devices for stem cell transplantation. We knew that even after we'd selected for stem cells some tumor cells remained in the final collection. Transplant doctors wanted a secondary system that would pick out those tumor cells before they were given back to the patient. It made tremendous sense, and we felt confident we could do it. But from a business standpoint the problem with that project was that it would take at least five years of clinical evaluation to determine its value. Possibly more. Logically, the fewer tumor cells you put back in a transplant patient the less chance he or she would have to relapse—that's obvious—but proving that to the satisfaction of the FDA would be extremely difficult. And if we couldn't prove it statistically we would not be able to get government permission to market it.

In order to conduct a fair test you have to begin with a level playing field; every patient has to be the same. That's impossible in oncology. Every cancer patient is unique. Patients respond differently to the same treatment. Cancer progresses at different rates in different patients. There is no standard treatment against which improvements might be measured. Treatment regimens are tailored to fit each patient; it's an art as well as a science. And it makes comparative studies very difficult.

Still, transplanters wanted a tumor-cell depletion column. Proof of value or not, they wanted to eliminate as many tumor cells as possible. The question was whether to focus on breast cancer or lymphoma.

Breast cancer is the second leading cause of death among American women. In the quest to find a cure, bone marrow transplants had become an accepted, though controversial, treatment for advanced cases. Ron Berenson wanted to concentrate on breast cancer because the initial studies seemed to indicate that tumor cell depletion might have a more immediate benefit in breast cancer patients than lymphoma patients. I agreed with him—because the opportunity was bigger and it looked like a more straightforward project. Results would be easier to measure.

This was a business decision for me and CellPro. Put simply, focusing on the breast cancer depletion project made good business sense. We could get to where we wanted to go faster. I knew that patients would be affected by whatever decisions we made, but we couldn't worry about that. We could help the most people with the breast cancer depletion project, and so I agreed with Ron that that's what we should pursue. Of course, three years later, when my own life was at stake, I desperately wanted that lymphoma B-cell depletion column. And then I was determined to get it.

So, with great confidence we launched a breast cancer depletion column project.

It proved to be far more difficult than we had anticipated. We had great difficulty finding good breast cancer tumor-cell antibodies, and our assays, our methods of analyzing our results, weren't sensitive enough to provide the kind of data we needed. So the project bumped along very slowly as we searched for better antibodies and tried to develop more sensitive tests.

Once it became clear that it would take us longer than we anticipated to develop a depletion system, we shifted priorities and began concentrating on projects that would improve our original Ceprate system. Fortunately, we had a lot of promising projects.

Much like a new video game entrances kids, purified stem cells excited medical researchers. They began to work with us on an incredible variety of exciting projects made possible by the adaptability of the Ceprate platform. By the summer of 1993 we were involved in thirty-eight different clinical trials involving hundreds of patients. A clinical trial is simply a controlled experiment. After a drug or treatment has been proved in a laboratory to be safe, it is used to treat a small group of patients. The result of this trial is compared to traditional treatments. If it is shown to have benefits without causing any harm, it will be approved for wider trials.

In trials in the United States and Germany doctors began treating patients with stem cells extracted from peripheral blood, circulating blood, by the Ceprate system rather than stem cells gathered by pounding long needles into bones hundreds of times and wrenching out the marrow. At most sites patients who received cells extracted from peripheral blood engrafted—reconstituted their immune systems—more rapidly than those who received whole marrow.

Gene therapy specialists believed purified stem cells could be an important new weapon against genetic diseases. Until this time they had been using white blood cells to carry modified genes into the body. Unfortunately white blood cells did not survive very long and did not reproduce. They envisioned stem cells as the bus that would carry their scientifically engineered genes into the body to effect permanent change. In March 1992 the National Heart, Lung, and Blood Institute at the National Institutes of Health began using stem cells to treat children suffering from SCIDS, a fatal disease caused by a missing gene. At the same time researchers at the University of Texas began inserting a genetic marker into stem cells, the first step in a program that would hopefully lead to the insertion of any desired gene—genes that would fight diseases, increase tolerance to cancer drugs, and even replace damaged genes. It was a whole new world.

At Johns Hopkins, Dr. Stephen Noga began transplanting a suspension of donated stem cells from which the T lymphocytes had been removed into patients suffering from acute leukemia, lymphoma, and multiple myeloma. T lymphocytes are immune system cells from a donor that attack a transplant recipient's tissues and organs, causing the often fatal graft-versus-host disease, or GvHD. Just as we planned to remove tumor cells from lymphoma patients before giving them back their cells, Dr. Noga removed these T-cells. In Dr. Noga's study patients receiving purified stem cells recovered more rapidly than those receiving bone marrow—and there was significantly less GvHD.

The potential applications for the use of purified stem cells in therapeutic situations seemed limitless: lymphoma, leukemia, breast cancer, multiple myeloma, AIDS, SCIDS, genetic diseases, multiple sclerosis, GvHD in solid organ transplants, small-cell lung cancer, T-cell depletion in allogeneic (donor) stem cell transplants, even diagnostic applications. We were the cutting edge of medicine. We worked with the optimism of the Spanish fleet, sailing in all directions to discover new worlds. So while I remained concerned about our legal problems, I managed to keep them in perspective. Most companies have legal entanglements that need to be unraveled and I continued to believe that somehow, some way, we would reach an agreement with Baxter. The alternative, that all the promise of CellPro would disappear in a legal dispute, was simply unthinkable.

The board's search for a CEO ended in June. I had been invited to speak at an investors' conference in Baltimore. Joe Lacob was also there, as were many other venture capitalists, to learn what was going on with the new companies. The scheduled activity for the evening was to attend an Orioles baseball game at Camden Yards. As Joe and I were walking over to the ballpark he said somewhat critically that I could have done a better job on my presentation. My talk had focused too extensively on the technical aspects of our system rather than our marketing prospects. I still didn't sell our product hard enough. "You really need to work on that more," he suggested, and then he added casually, "But we've decided that we're going to appoint you CEO of the company."

I wasn't completely surprised, but I was thrilled. I just felt wonderful. There is a great satisfaction that comes from being in command, whether it's of a small craft or a big company. I called Patty as soon as I got back to the hotel. Patty is not someone who responds to good news or bad news with great emotion, but she was as pleased as I was. There had been a lot of moves in our life together to finally reach this moment.

Each person feels quite differently about the company for which he or she works; some people love what they do and where they do it, for others a job is simply a paycheck and they have no emotional investment in their employer. They do their work and they go home. My strongest ties had been to the small companies. For example, I had been a part of HemaScience since it consisted of a rough prototype on a workbench in a garage. I was extremely proud of what we'd built. HemaScience was, at least partially, my company and I loved it. Baxter was a multibillion-dollar corporation before I got there, and I knew I would never make a significant difference. While I enjoyed many things about working there, particularly some of the people with whom I worked, I never really felt a part of the fabric of the company.

CellPro was still small enough when I began working there for me to affect its future. Unlike when I worked at Baxter, decisions I made one week would be instituted the next week. It surprised me how quickly I developed a real relationship with this company. I cared a lot about it on a very personal level very quickly. I think everyone at Cell-Pro realized that we were all playing significant roles in a potentially very important and very successful young company. We were small enough that I got to know just about every one of our employees on a personal basis. I went to the picnics and the social events, I wandered through our laboratories.

We were, all of us, confident, proud and damn good. CellPro was a fine place to work.

I'd been in the medical industry for two decades and had learned a lot while doing a variety of jobs at very different companies. I'd never read any management books nor had I attended many executive seminars, but I had developed my own theories about how best to run a business and finally I had the opportunity to put them into practice.

In addition to creating the corporate culture, there are three things a CEO has to do well: make sure the company is well financed, hire the best people possible and put them in a position to do those things they do best, and make sure the corporate resources are being used most efficiently.

Finding good people was a continuous process; I didn't always pick the right people. Nobody can. Sometimes I hired people who didn't work out and within a few months I had to fire them. Fortunately our CFO, Larry Culver, was very good at bringing in talented people, and took a tremendous amount of pressure off me.

When I'd first met Larry Culver I had been put off by him. He seemed cold and distant; I didn't think he liked me very much. I probably had even identified him as a potential problem for me if I ended up at CellPro. It would be difficult for me to have been any more wrong about someone. As soon as I started working there Larry Culver became my friend and advisor. Like just about everyone else at CellPro, Larry loved being part of a company whose business was helping save lives. We ate lunch together every day for years. It became such a habit that people used to make jokes about it. But in a battle this was the man you wanted in your foxhole.

Among the first people I hired was Dr. Joseph Tarnowski as our Vice-President of Operations and Chief Technical Officer. CellPro had successfully gotten through the most difficult period for any young company—survival—and had become reasonably well financed with a strong technology base. Joe was in charge of manufacturing, development, and quality control, which involved overseeing the entire process from feasibility to commercialization, every stage from "Okay, let's do it" to "Does this come with a warranty?" This was the job where science met engineering, chaperoned by endless government regulations; it was the job with a million details and an equal number of headaches.

Tarnowski very quickly became our joe-of-all-trades. He had previously worked as a bench scientist, the researcher who mixes the chemicals, at the prestigious Roche Institute of Molecular Biology. When the pharmaceutical giant Hoffman–La Roche began developing the drug interferon, which for a time was believed to be the cure for cancer, Tarnowski was hired to purify proteins. As senior managers started leaving that company to dive into the new world of biotech, Tarnowski raced up the ladder. He ended up doing all sorts of things senior scientists rarely get to do, becoming involved with all aspects of product development from lab management to government regulatory issues. Eventually he was recruited by a biotech start-up desperate for someone with experience in product development, manufacturing, facilities management, and plant development. In other words, everything. That was pretty much his job description at CellPro.

CellPro never made a profit selling product, and there was never a single day when our revenues approached our expenses. It was Wall Street's faith and one shortlived strategic partnership that kept us in business and enabled many investors to make a substantial profit. Toward the end of 1992 Larry Culver and I decided to do another round of financing. A well-run company doesn't wait until it needs money to try to raise it. When market conditions are good to raise money, you do it.

We thought we were very well positioned when we announced this secondary offering. Our stock was selling at almost $24 a share, the entire biotech sector was performing well in the market, we were in Phase III clinical trials of the Ceprate system and expected to be on the market within a year, and we were getting positive results in other early tests. We had a good story to tell. We anticipated raising as much as $65 million with this offering.

As it turned out, we sold into a down market. The bloom was off biotech. Important companies had announced disappointing results. Wall Street had just picked up the scent of information companies. Brokers were hearing new words like "Internet," "software," and "Web site." The entire biotech sector was being hammered when we sold, and as a result we were forced to sell our shares at $16.50, considerably less than we had expected. Although investors were still betting on us—the week after our offering CellPro was one of the ten most actively traded stocks on the NASDAQ exchange—we raised "only" $38 million. We had $55 million in the bank before we'd even gotten FDA approval of our Ceprate system. But we felt certain that getting approval was only a matter of time.

The other potential source of revenue was a corporate partnership. We had received approval to begin selling the Ceprate system in Europe. Although we had a small European marketing operation, we needed to find an international partner willing to make a major investment in Cell-Pro in return for the potentially very valuable rights to sell our products outside the United States. In fact, one of the reasons I'd originally been hired was to find a company with whom it made sense to form a strategic alliance. About the only thing we now knew for certain was that our partner would not be Baxter.

Eventually we reached an agreement with a relatively staid $3.2 billion family-run company named Corange International Ltd., the parent company of Boehringer Mannheim. Corange was the second largest manufacturer of diagnostic tests and biochemical products in the world. After only a few weeks of negotiations Corange made us an offer valued at $220 million. I remember sitting in a room with Larry Culver listening to this proposal being outlined. Negotiating is simply high stakes poker. And as in poker it's vital not to give away your hand with your emotions. So I was particularly pleased that when they made that proposal I sat relatively stone-faced, nodding occasionally at the details, rather than leaping up, thrusting my fist into the air, and screaming "Yes!" Instead I responded calmly, "Please give us a little time to consider this."

Larry Culver and I went into another room and Larry told me, in his typical understated way, "Well, I think I'd have to advise you that this is a pretty reasonable deal for us."

It was an extraordinary deal. Unbelievable. If there was any question that CellPro was the most important company in the cell separation business this deal ended it. It was the largest nonacquisition biotech deal ever done. Basically, Corange bought 15 percent of CellPro at a price that valued the company at $730 million. The *Wall Street Journal* reported that this "eye-popping figure" for a company with annual revenues of $2.5 million, "clearly anoints [CellPro] as the leader in the high stakes race to develop systems known as cell separators."

This deal justified completely our refusal to accept Baxter's terms. If we had given Baxter what they had demanded, nonexclusive distribution rights to the United States and exclusive rights to Europe and Japan, the Corange deal would not have been possible.

When this deal was announced our stock soared to $35 a share, the highest it had ever been. It also made me a very hot commodity. Concluding a corporate partnership with a respected multinational company like Corange at these numbers earned me a lot of attention. Headhunters began calling to ask if I was interested in exploring other possibilities. Financial reporters called to discuss biotech matters. It was all very flattering.

Patty and the boys didn't let this attention affect me. Patty made it clear it was still my job to take out the garbage Tuesday nights. And the boys still didn't let me select the movie for our family get-togethers— although it remained my responsibility to make the popcorn. My family had settled nicely into Seattle. We'd brought our boat up from San Diego, and the boys and I often spent weekends sailing together. Jamie and Ben were becoming fine sailors, and my long-time crewmate, Patty, no longer felt compelled to sail with us on every trip.

While I was often overwhelmed with CellPro business, I tried hard to incorporate it into my life, rather than making my work my life. If in the heat of corporate battles I sometimes forgot why I was working so hard, Patty was always there to remind me. We didn't just love each other as relatives, we actually enjoyed spending time together. I've always loved opera, for example, and I was pleased to discover that Ben did too. We had season tickets to the Seattle opera, with good seats. The two of us would put a bottle of champagne on ice in the back of the car and drive into the city for the performance. At intermission we'd go outside and have own little "opera" tailgate party. Wherever we've lived we've developed Murdock family rituals. When I was at HemaScience, in Orange County, we discovered that late in the evening of December 23 the crowds at Disneyland disappeared. For about three hours the park seemed to belong to us. There were no lines at any of the rides. So for those three magical hours we'd run ragged through the entire park. We'd ride Space Mountain twice, the Matterhorn, any ride we wanted. Going to Disneyland on December 23 became an annual event until we moved to Belgium.

To my delight, both of the boys became interested in the business of science. Jamie and Ben spent several summers working, for free, as interns at CellPro. We had become a CellPro family.

The Corange deal was so exciting it allowed me to temporarily forget that the patent lawsuit remained a very serious problem. The lawsuit was moving inexorably toward trial, although I still much preferred to negotiate a settlement. It just seemed impossible to me that there wasn't some way of resolving this dispute. Deals much more complicated than this are completed every day. It would have to be a straight license deal however; we would pay fees and royalties, because the Corange deal had made it impossible for us to offer Baxter any kind of distribution arrangement.

With negotiations at an impasse Baxter agreed to join us in nonbinding mediation. The hope was that a neutral third party could beat both companies into a fair agreement. That requires a talented mediator—and the mediator turned out to be little more than an expensive errand boy. A mediator will usually beat up one side, then beat up the other side and convince both of them that they're in poor negotiating positions and push them toward the center. This mediator's main contribution was getting us sandwiches and drinks. He ran back and forth between meeting rooms, telling each of us what the other party was offering. He made very little attempt to identify a common ground and try to expand it.

It was roughly during this same time that negotiators were trying to work out a peace plan for the Middle East. Incredibly, they had more success bringing together the Israelis and the Palestinians than CellPro and Baxter.

The mediation was a disaster. We were attempting to negotiate a reasonable royalty rate, based on the fact that we had made stem cell transplantation a reality. We were willing to pay them a toll, but we were not willing to give them the farm. They responded by raising the price of a settlement. They had decided that their patents had become so much more valuable—because of the work we'd done—that the up-front fee had become at least "double digit millions." Basically, the price of our success was now ten times what Baxter had originally requested.

At the end of the second day, after it had become obvious we were not going to be able to reach an agreement, they filed an already prepared lawsuit against us in Delaware claiming patent infringement. So it was difficult for me to believe that they had come to these negotiations in good faith.

When we had filed our lawsuit we did not include Johns Hopkins as a defendant. We did not want to sue one of our most important clinical sites. The District Court in Seattle decided that Johns Hopkins, as owner of the Civin patents, was an indispensable party to this case. The judge determined that the court had no jurisdiction over Hopkins because the university didn't have an office, any property, or an organization in Washington. Legally, Johns Hopkins did not exist in the state. The judge didn't buy our argument that the court had the right to summon Johns Hopkins because it raised funds there. So when Baxter, BD, and Hopkins filed their lawsuit, the judge in Seattle allowed the lawsuits to be consolidated in Delaware. We'd lost the home court advantage.

Delaware may be a small state, but in the world of corporate legal affairs it is dominant. Early in the 1900s Delaware's low franchise tax and very pro-business legal codes made it an extremely attractive place for companies to incorporate. For legal and tax reasons almost 60 percent of the *Fortune* 500, America's largest companies, are incorporated in Delaware and operate under that state's laws. Big business is big business for Delaware, the corporate taxes and fees these companies pay account for approximately 20 percent of the state's annual operating budget. Even CellPro was incorporated there—although our actual physical presence was limited to a file and a corporate stamp in a file cabinet. Like most other companies we paid a registered agent $1200 a year to act as our representative for certain legal affairs.

Because almost all of the companies registered there are actually located in other states, for corporate lawsuits Delaware is considered neutral territory. It's the O.K. Corral of big-time business litigation. The Delaware judiciary has developed a reputation for expertise in corporate law, particularly patent disputes.

There was to be one final attempt to negotiate a settlement. Several months later, as part of the pre-trial process a magistrate chaired a settlement conference. Judge Trostle was tough; she pushed hard on both sides, but it was obvious we were not making progress. The judge was a heavy smoker, and Kiley joined in, so after several hours the room was enveloped in a blue haze. I got a terrible headache, my eyes were burning. But toward the end of the day she told us, "I have an unusual request from the other side. John Osth would like to meet alone with Rick Murdock."

Kiley bristled at the suggestion. He was our lead negotiator and did not like being replaced at all, but I felt we might be able to make some progress. I had known John Osth at Baxter. I liked him. And we had absolutely nothing to lose. "I'm not going to give anything away," I promised Kiley, "but what harm does it do to talk to the guy?"

Osth and I met with Judge Trostle. He seemed very conciliatory. "Look," he said, "how did this thing go so wrong?"

I sighed. "Boy, John, I don't know. I can tell you that none of us wanted this to happen, but we didn't feel we had any place to go and we had to protect ourselves. We've always wanted to solve this and we'd still like to solve it now."

Baxter wanted to solve it too, he responded. He asked me to outline various ways our two companies might work together, "instead of slugging it out in court," and then we would meet again to try to hammer out an arrangement.

I did my homework. Working with several of my people we identified several areas in which we might profitably cooperate with Baxter. For example, they had a spectacular facility for manufacturing monoclonal antibodies, and we proposed that they manufacture our antibody for us. We were willing to discuss manufacturing other components of our disposable kit, the "software" portion of the Ceprate system. There were even some ways we could work together in Europe on products not covered by the Corange deal. I thought we'd made a real effort to get this out of the confrontational mode into the collaboration-partnership mode.

In September 1994, I met with John Osth at the Biotech CEO Conference at Laguna Nigel and presented him with our proposal. I felt very good about this meeting; this was two guys who respected each other trying to make a reasonable deal. We were doing business, not shouting at each other through lawyers. I put some meaningful proposals on the table and Osth seemed receptive. It seemed as if we might be able to find a way out of this mess. Osth received my suggestions enthusiastically. "We've got some things here that we can do to solve this," he said, "I really believe we can." I left that meeting feeling a twinge of optimism.

We never heard from Baxter again. It was astounding. They never responded in any way to my proposals. But by that point I believe that people above John Osth had just decided to bury us. The next time I saw or heard John Osth's voice was in the courtroom. What happened wasn't his fault; the decisions were being made in Deerfield, and he was simply a pawn.

We'd spent more than two years trying unsuccessfully to make a reasonable deal with Baxter. Baxter claims that the inability to do that was caused by our intransigence, but I believe that they never wanted to make a fair deal. We had proved that this market was potentially worth hundreds of millions of dollars and they intended to dominate it. What they couldn't do in their laboratories or win at the negotiating table they expected to get in the courtroom.

I began preparing myself for a brutal legal battle. Baxter played for keeps. I was ready for that. But I wasn't ready for everything else that hit us at that time. Suddenly, the foundation that we had so carefully built at CellPro began collapsing. Nothing I had experienced or learned prepared me for the sea change that was about to take place in my life.

4 I WAS READY FOR THE TRIAL TO BEGIN. WHILE I remained incredulous that we had not been able to work out some kind of deal, I was confident that we would prevail in court. Didn't the good guys always win in the end? Wasn't right stronger than might? Although, somewhere in the recesses of my mind, I was always aware that we could lose, I just didn't believe that was possible; we were using our own antibody in our own stem-cell selection technology. At times I tried to focus on the only positive aspect of the trial; when it was over, CellPro would no longer have the dark shadow of Baxter hanging over it.

Josh Green had complained that our defense lacked focus, that it seemed as if we were throwing everything we had against the wall in hopes that something would stick. Maybe that was true. Josh Green is a very smart man. I didn't understand all the legal maneuvering, though I had tremendous confidence in Coe Bloomberg.

I had spent a lot of time preparing for the trial. It took large chunks out of every day. I worked closely with our lawyers, preparing to be deposed by their lawyers—then I spent several days being deposed by Baxter's lead counsel, Don Ware. During a deposition, I had been well taught, there was no such thing as an innocent question. Every question had some deeper meaning, some connection to a hidden strategy. So I found something suspicious in every question I was asked. I didn't quite know what their strategy was, but I was careful not to give answers that would help them develop it. Before answering even the simplest question, I learned to pause and examine it, turning it over in my mind like a computer examining an interesting shape, searching to see if it had the makings of a trap that months later might suddenly be sprung on me in the courtroom. "But, Mr. Murdock," I could imagine Ware thundering in front of the jury, "didn't you say very clearly in your deposition that . . ."

I kept reminding myself this wasn't a game, but the more I became involved in it, I realized that that was exactly what it was, a complex, high-stakes game.

We hired an array of experts to help us win. I'm sure Baxter, Becton Dickinson, and Johns Hopkins did too. We even hired a jury consultant to help us select jurors from the larger jury pool who would be most sympathetic to our case. This was a very expensive game.

I also helped our lawyers prepare to depose Baxter's witnesses. I sat in on every strategy session. As we got closer to the beginning of the trial, I sometimes felt as though I were sailing in a thick fog bank. The trial completely dominated my world; all the landmarks by which I usually navigated had absolutely no application. Like a fog, the trial enveloped everything, it was impossible to see beyond it. But worst of all was the feeling that I'd lost control.

I approached the trial with a fear of the unknown. A feeling that there was something else out there, something beyond my sight, something very dangerous.

I was absolutely right.

I've never paused to look for a reason why things happen. I'm not a superstitious person. I don't believe in astrology. I don't believe in karma. Although I attended a religious school, I'm not a particularly religious person. I don't even believe in luck. So I don't think there is any explanation, logical or fantastic, for the way my life began changing both professionally and personally at this time.

In addition to the trial, three significant problems hit us almost simultaneously. If not solved, any one of them could literally destroy the company. First, once again an internal battle for control of the company had broken out. Second, six months after buying 15 percent of CellPro, the CEO at Corange was forced to resign, putting that entire deal in jeopardy. And, third, in late December 1994 we finally got a response from the Food and Drug Administration about our pending pre-market approval application for the Ceprate system. We had closed the company for the week between Christmas and New Year's. But when Ron Berenson stopped in to pick up the mail, he found a formal letter from the FDA. As he read the letter, he was staggered.

I was working out on my NordicTrack when he called. "Ron's on the phone," Patty said. "He says it's really important."

"Hey, buddy," I said happily, "what's up?"

I could hear the panic in his voice. That wasn't like Ron at all. "Rick, you're not gonna believe this. We just got a letter from the FDA. They said there's a problem with the clinical data. This thing is like ten pages long and there's all kinds of questions and issues." He paused, then almost whispered, "This is unbelievable."

I sat down in shock and took a long, deep breath, trying to gather my thoughts. "I'll be right there," I said. As I read the letter, it seemed even worse than Berenson had told me. The FDA was informing us that the Ceprate system was not going to be approved. Based on the data we'd submitted for approval, it appeared that stem-cell purification was actually harming patients. I immediately called Larry Culver. I suspect he heard the same panic in my voice that I had heard during Ron's call. Larry was there very quickly, and the three of us began planning strategy. As a public company we had to disclose the contents of this letter to our shareholders. It was potentially devastating to our stock. But as I read the letter over and over, it just didn't make sense to me.

The Ceprate system fit into the traditional FDA regulatory approval structure about as well as a jazz saxophonist fits into a string quartet. For product evaluation, the FDA is divided into three groups: Drugs, Biologics, and Devices. We didn't fit neatly into any of them. We certainly weren't a drug; our biological component, the monoclonal antibodies, was not active within the patient; and, unlike most other therapeutic medical devices, the Ceprate system had no physical contact with the patient. The fact is, the FDA really didn't know how to regulate our system.

The Food and Drug Administration moves very slowly and tends to be conservative in its decisions, based—properly—on the theory that it's better to delay approval of a good drug than permit a bad drug to be marketed. This is a life-and-death business. If the FDA makes a mistake, people may die.

The FDA had complete regulatory control over CellPro. Everything we did was subject to its approval. It really would determine whether we survived and prospered or closed the doors on this new technology. But there was good reason it had been given this awesome power. There was a time, less than a century ago, when most foods and drugs were unregulated. Long before late-night TV existed, anyone with a wagon, some bottles, and a good spiel could sell "snake oil" to the public. These "humbug" salesmen didn't have to prove their product was either safe or effective; the sole legal requirement necessary to avoid prosecution was that they personally had to believe this remedy worked. In many cases the only thing between the cures these people were selling for just about every disease imaginable and the gullible public was the local sheriff. While most of these products were about as harmless as the "lose weight while you sleep!" infomercials, some of them were dangerous.

The agency that would eventually become the FDA was founded in 1862, when President Lincoln appointed a chemist to the newly formed Department of Agriculture. This led to the establishment of the Bureau of Chemistry, which in 1927 became the Food, Drug and Insecticide Administration, and three years later the Food and Drug Administration (FDA), which had limited powers. That changed in 1938, when the Massengill Company began selling a liquid form of sulfanilamide, a drug commonly used in powder or pill form to treat streptococcal infections, mostly sore throats in children. Unfortunately, the liquid in which the drug was dissolved was diethylene glycol—a deadly poison commonly known as antifreeze. One hundred seven people, most of them children, suffered horribly for weeks before dying. Incredibly, the fact that Massengill had marketed a poison did not violate any existing laws. In response to public outrage, Congress passed the Federal Food, Drug and Cosmetic Act of 1938, which for the first time required pharmaceutical companies to prove their products were safe before they could sell them—although they still were not required to prove they actually did any good. It also extended the reach of the FDA to cosmetics and therapeutic devices.

It was only in 1962 that Congress passed legislation requiring drug companies to prove the effectiveness of their products before they were permitted to market them. Ten years later, in 1972, the FDA was given the additional responsibility of regulating biologics, including blood products. And the Medical Device Amendment, passed in 1976 after stories concerning faulty IUDs and pacemakers were printed, gave the FDA jurisdiction over medical devices—including our Ceprate system.

At the end of 1994 we had been waiting expectantly for official approval to begin marketing our system. Three years of clinical trials had gone very well. The doctors who used the system were enthusiastic about it. So we were stunned when the FDA turned us down.

According to FDA investigators, the data we'd submitted indicated that patients receiving purified stem cells had taken longer to regenerate platelets, the blood cells responsible for blood clotting, than patients receiving whole marrow. That meant that some of these transplant patients might remain at risk of bleeding for a longer period.

We were aware of the small difference in time, but our investigators considered it to be of no clinical significance. The data showed what the FDA claimed, even though patients had experienced absolutely no problems in more than sixty clinical trials. We wondered if the real problem might be with the data rather than with the system. I put the best possible spin on it, but I didn't fool anybody. The day we did so, our stock fell almost 25 percent. Wall Street warned that a lengthy delay in reaching the market would allow competitors to catch up with us. Without FDA approval, we were out of business.

In addition to dealing with the lawsuit and the FDA, once again I had to resolve internal problems. The struggle between scientific development and marketing had flared once again. There is always an uneasy truce between scientists and salesmen. Scientists want their product to be absolutely perfect before allowing it to be sold; salesmen want to get it out the door where it can start generating income as soon as possible. On occasion the truce is broken, usually over budgetary and resource issues.

Unlike the situation several years earlier, by this time Berenson's technology had matured into a sophisticated device and our emphasis had changed. Marketing the system, getting it in the hands of doctors, and discovering its capabilities when used in research and patient treatment had become our priority. We were a public company now, and on this swing of the pendulum the business side held the power.

Ron Berenson was a brilliant scientist who had invented a cell-separation technology that had the potential for saving thousands of lives. He deserved tremendous credit for bringing the company he had founded to the point at which he was no longer essential. Realizing that he wasn't going to win the battle against the business side, he submitted his resignation, which was accepted. That was tough to do. Ron Berenson and I had spent a great deal of time together. We'd traveled around the world, we'd shared victories and defeats. In addition to respecting him, I also liked him a lot. But the day he resigned was the last day we ever spoke. Our stock dropped at the news that CellPro's founder had left the company, but eventually recovered.

The Corange deal was doomed from the start. When Max Link, the CEO with whom we had made our deal, left, none of the remaining executives had the slightest idea what to do with us. This deal could have been extremely profitable for them—if they had understood the new world of stem cells. But they didn't have a clue.

It became obvious that they were looking for a way to get out of the contract. In April 1995 they held up a scheduled $60-million payment and announced they were reconsidering the deal. The truth is, we wanted out of the deal, too. We were locked into a worldwide distribution deal with a corporate partner who lacked the desire or understanding to market our technology. The question was how best to end the arrangement. Corange simply wanted to walk away. As their new CEO told me, "We've decided we don't want to be in oncology anymore."

I reminded him, somewhat forcefully, that we had an ironclad agreement and that we were prepared to implement the deal under those terms. I think I might have said, "I don't want to get into another lawsuit, but you guys owe us a lot of money." Get involved in another lawsuit? The absolute last thing I wanted to do was get involved in another lawsuit, but I had a legal obligation to our stockholders.

We met with the Corange people in New York to try to dissolve this partnership peacefully. There was a lot of dissension within our group about exactly what terms we would accept. Kiley wanted to go to the wall, make them pay every penny or sue them. Same old Tom. Testosterone coming out of his ears. "You know something?" I said. "For $30 million I'd walk away from this deal in a microsecond."

Tom thought about that for a moment. "Thirty million? That's the number?"

"Yeah," I said, nodding, "I'd walk for that."

Kiley and Josh Green met with Corange in a lawyer's office and played "good cop/bad cop." Tom was the bad cop, and walked out of the room with $30 million to terminate the agreement.

It turned out to be a very good deal for us. They had paid us a total of $90 million for a 15-percent stake in the company, and we had regained our worldwide marketing rights. Investors thought it was a good deal, too. When the agreement was announced, our stock immediately rose two dollars a share.

But the problem with the FDA could not be solved through negotiation. To the FDA, this was a serious safety issue. Our own data had shown a problem. At the beginning of Joe Tarnowski's career, he had been shown a training film outlining the basic FDA rules and regulations concerning a manufacturing environment. The "sad dog film," as he described it, showed a happy dog eating his food—then rolling over and dying. Food, dead dog, film over. It made an impact. When the lights went on, Joe remembers, people in the room were crying. The film's narrator explained that a small problem in the manufacturing process had allowed a piece of shrapnel to fall into a can of dog food and end up in the dog's stomach. Joe never forgot it—and it served as a reminder that the FDA isn't playing games or politics.

In February 1995 we met with FDA representatives who told us, in the complex language of bureaucracy, "Based on the data you've submitted, you don't have a snowball's chance in hell of getting this thing approved." The FDA's own team of scientists and statisticians had evaluated our submission. Their analysis of our data indicated that purified stem cells led to a two-day delay in platelet engraftment, which meant patients faced the risk of uncontrollable bleeding for two additional days. If the FDA analysis was correct, this was a safety problem that the entire stem-cell selection industry would have to deal with.

This was a potentially devastating discovery. If purified stem-cell transplants weren't safe and efficacious, we didn't have much of a business. It appeared that we would have to start a new round of clinical trials, and it would be at least another five years—if ever—before we got FDA approval. We were burning a lot of money turning on the lights every day; if we didn't get the Ceprate system on the market pretty soon, there was little chance we'd be able to stay in business long enough to complete another round of clinical trials.

But when we began investigating their complaint, we were confused by the fact that our clinical experience just didn't support the statistical data. The story the numbers were telling was very different from what doctors had experienced. We hadn't received a single warning from doctors who had actually used the Ceprate system in transplants. Not only weren't our field investigators concerned about this problem—it appeared they didn't even know it existed. Dr. Cindy Jacobs, our Vice-President of Clinical Research, decided to plow through all the numbers again to see if she might make sense of them. "There's something strange going on here, Rick," she told me. "If you give me some time, I can solve this."

Cindy Jacobs was the best clinical regulatory director with whom I had ever worked. She was an M.D./Ph.D. from Montana, a woman who loved being outdoors just as much as she loved crunching numbers. Cindy lived with her husband, who was a landscaper, and their child in a beautiful wood-beamed ranch house they'd built out in the woods. Patty and I had been there for parties several times. After Berenson's departure, Cindy had assumed a lot of responsibility, and I had come to rely on her to produce everyday miracles. So if she told me she could solve this problem, I believed her. But this was an unusually difficult problem. Numbers don't lie.

Modern science is defined by numbers. An experiment is a mathematical equation whose result is expressed in numbers as well as words, and the numbers describe it more precisely than the most exact spoken languages. The numbers tell the truth over centuries: Antoine Lavoisier, the founder of modern chemistry, who died in 1794, can still speak clearly to Ron Berenson or Cindy Jacobs through the numbers. The numbers may seem to be cold and unfeeling, but to those people who understand them, they are as rich with emotion as Shakespearean sonnets, causing joy or sadness, hope or frustration.

Cindy heard the music in the numbers. She took three statisticians with her off-site, and for several days they pored through our data. Hidden beneath that mountain of statistics that had been generated in clinical trials, Cindy and her team discovered the reason for the delay in platelet engraftment. The answer seemed so obvious—but we were exploring a new universe. We were establishing baselines, learning literally day by day how the body responded to purified stem cells. We had no basis for comparison. We didn't even know what was right, so it was difficult to pick up what was wrong. As Cindy proved, the delay in platelet engraftment was directly related to the number of stem cells the patient received. It was so simple: If patients didn't get enough stem cells back, they occasionally took a longer time to recover. As the numbers told her, there was a threshold of 1.2×10^6, or 1.2 million, cells. Our data showed that if fewer than that number of stem cells were returned to the patient, there could be a delay in platelet engraftment, but if more than that number were returned, there was no delay. We proposed to the FDA that we solve that problem by putting a warning on the label.

The numbers made sense to us, but we weren't sure what impact they would have on the FDA. While we waited for the FDA's response, we continued to provide the Ceprate system for clinical trials and research on a cost-recovery basis. We could charge customers what it cost us to produce the system. With the engraftment problem solved we hoped that our application would soon be approved. But many months passed and we didn't hear from the FDA. This had nothing to do with good science or bad science, product safety, or public benefit, this was just paperwork. Too many products, too few people to regulate them. Because of this, a lot of promising technologies languished in the bureaucracy for a long time.

One by one we were solving our problems: we were back on track to receive FDA approval for our product. The deal with Corange had been concluded with a substantial payment—and we had the distribution rights back. The internal problems had been resolved. There was only one more hurdle to clear—although at times this hurdle often seemed to be about as high as the Empire State Building.

The trial began in Wilmington, Delaware, at the end of July 1995. Several weeks earlier our attorneys had conducted a mock trial to determine the strengths and weaknesses of our case. Even though it went fairly well, Josh Green had remained a harsh critic of the decision to fight it out in court. After listening to our arguments for several hours at a CellPro Board of Directors meeting, he threw up his hands in frustration and complained he still didn't understand our case. "How is a housewife on the jury going to understand this if I can't?"

Kiley insisted we had a strong case. "You're a corporate guy," he responded to Green, "you're not a litigator. That's why you don't understand."

"I don't understand because it's not clear," Green said. "We're all over the place. I have got to tell you, if there's any way we can settle this thing, we ought to try to do it right now, 'cause I'm not comfortable with our case at all."

"We don't have that option," Kiley reminded him. The tension between Green and Kiley finally erupted, and they started yelling at each other across a highly polished conference table. Maybe I didn't know much about the legal system, but the sight of my two lawyers shouting at each other was not a big confidence-builder.

I didn't know what to think. Both men made sense to me. I just didn't know enough about the judicial system to know if our case was as weak as Josh Green claimed or as strong as Tom Kiley insisted. I certainly knew every single fact, and I believed without any doubt that we were right, but, as I was just beginning to understand, in a court of law being right is often not enough. In my mind I envisioned this as a case in which we were the good guys in the white hats, building a profitable company that was eventually going to save countless lives, and the other side were the outlaws wearing black hats, who cared mostly about robbing the bank.

It was as hot as hell in Wilmington, Delaware, that summer. I'd brought enough clothes for several weeks, although Judge McKelvie had made it clear this case would be concluded in two weeks, as he had scheduled his vacation and didn't intend to miss it. What surprised me most was how many lawyers were involved in this case. I sat at our counsel's table with our two attorneys from Lyon & Lyon, Coe Bloomberg and Bob Weiss. The three parties opposing us were represented by three separate law firms with so many attorneys that there wasn't enough room for all of them at their table.

Oddly enough, the jury consisted of seven women. In civil trials, I learned, the traditional twelve jurors were not required. It was obvious these jurors were trying very hard to understand the often complex scientific issues being discussed, but the language of science does not translate easily into terms comprehensible to the layperson.

Sometimes, when the testimony was particularly complicated, I'd glance at the jury expecting to see one or more of them lost in reverie. That never happened. Their attention never wavered. These people took this job very seriously; several of them took copious notes. But as Bob Weiss had explained to me one day, "They're not going to be able to understand everything that takes place. That doesn't really matter. In the end they're going to make their decision because they believe one side is telling the truth and the other side is not. That's what it's going to come down to, that feeling."

Baxter's argument, basically, was that it had purchased an exclusive license from Becton Dickenson, who had purchased it from Johns Hopkins, to four patents that had been properly issued by the government of the United States, and that we were infringing those patents: the first, known as the 204 patent, by using our antibody to select for stem cells, and the second patent, the 680, simply by creating a suspension containing more than 90 percent stem cells. There were two other use patents.

Our response was that these patents were invalid for several reasons, but even assuming they were valid, we were not infringing them. We stressed several points in our presentation to the jury: A patent is a recipe. It must include all the information a skilled researcher needs to replicate what was done by the inventor and achieve the same results. Meaning the meal has to come out of the oven the same way every time, no matter who cooks it. If a patent isn't "enabled," meaning the instructions don't allow someone else to achieve precisely the same results as the inventor, it isn't valid. It was obvious that the Civin patents didn't teach anyone how to cook that meal. Even Dr. Curt Civin, working in his own laboratory, was unable to find additional stem-cell antibodies by following his own recipe. If the recipe had been as easily followed as required by patent law, many other stem-cell antibodies should have been discovered quickly—especially in Civin's own lab. But they weren't.

We were also claiming that the 680 patent specifically protected a suspension of stem cells greater than 90-percent purity, and the Ceprate system did not produce a collection anywhere near that pure. We normally got between 50- and 70-percent purity. Ironically, we'd discovered that higher purity was not particularly desirable. Our investigators believed that in addition to stem cells, other cells could play an important role in engraftment, but no one knew exactly what role was played by which cells, and how many of them were desirable. So it was vital that a significant number of other cells be present. Civin's patent claimed 90-percent purity, we obtained 50 percent; it seemed quite clear we were not infringing that patent.

In addition to arguing that Civin's patent didn't enable skilled people to create stem-cell antibodies, and that our device didn't produce a 90-percent pure stem-cell collection, we also intended to prove that we were using an entirely different antibody. Baxter was entitled to protect the use of its My-10 antibody, nobody disagreed with that, but that didn't give them license to every other stem-cell antibody.

Our advocate was Coe Bloomberg, a large man of Norwegian descent. Coe was the quietest man I'd never known. He spoke only when he had something that needed to be said, and usually in response to a question. When I'd first met him, that had confused me. I'd try to draw him into a conversation, and instead of responding he would remain silent. For a time that had made me nervous; I didn't know what it meant. Long silences, I've learned, tend to make people uncomfortable. But as I got to know him well over the years when we were fighting this battle, I realized that that was his personality. His words were always well chosen.

Coe's strength was his ability to combine meticulous preparation with an agile mind; he knew all the facts, he understood the science, and he knew when to use that information. Watching him at work reminded me how majestic the law can be when practiced by someone with respect for its purpose.

I sat in the courtroom at the counsel table every day. There was not one moment during the entire trial that my heart was not racing. By that point I had built up a deep animosity toward Baxter. I had forged a strong emotional attachment to CellPro, and the people sitting only a few feet away from me were trying to destroy it. I had changed during the process. I didn't just want to win the case, I wanted to beat Baxter. In Europe, where the regulatory process is less stringent, our Ceprate system was already competing with their Isolex cell separation system—and with less than one-tenth the number of people and marketing resources, we were beating them handily. I wanted to go out into the domestic marketplace and bury them. It had really gotten nasty.

During that trial I gained great appreciation for what it takes to become a lawyer—and I developed a disdain for the entire legal process. I suppose I shouldn't have been surprised to learn that law as practiced in the courtroom has little to do with arriving at the truth. It's about winning. Nothing else matters but winning—nothing, whatever the cost in personal integrity. It isn't just a matter of splitting hairs with words, but splitting the cells that make up the hairs, then splitting the atoms that form the cells that constitute the hairs.

Each evening I would call home and describe in detail the events of the day. Ben, more than Jamie, who was getting ready to go to college, wanted to hear every bit of news. Finally, Coe Bloomberg suggested that Ben join me in Delaware for the remainder of the trial. Ben arrived and leaped willingly into that cauldron. And for him it became a defining experience. He stayed up most of every night working with our attorneys and paralegal staff in the war room, searching through piles of papers for specific documents, making copies, picking up and delivering. The next morning he'd put on his suit and spend the day sitting right behind me, intently watching the process.

The trial lasted two weeks. Because I had no experience in such matters, it was impossible for me even to guess how we were doing. Our lawyers reassured me that we were presenting a strong case. I was so naïve I believed them. As the days passed, it seemed to me that the judge, Roderick McKelvie, was against us. It wasn't blatant, but it seemed as if every important ruling favored Baxter. I accepted the fact that I was biased, so I dismissed my observation as the product of a slightly paranoid mind. But as I later learned, Larry Culver was reporting to Josh Green, "This judge seems like he's totally against us."

The hardest thing for me to do was simply sit quietly. At times I had to restrain myself. I remember squirming in my chair one afternoon as a Baxter executive swore to tell the whole truth and then proceeded to testify that Baxter had no business plans that related to stem-cell therapy. I knew with certainty those plans existed because I had seen them. I'd worked with the people who'd created them. But he insisted they did not exist.

I suspect the jury was swayed when Bloomberg caught one of Baxter's expert witnesses blatantly attempting to change his testimony. In response to a question regarding the definition of "substantially pure," this witness replied firmly, "Substantially pure means more than 50 percent."

Bloomberg immediately showed the jury a segment of this witness's videotaped deposition, in which he'd stated with the same confidence, "I'd say substantially pure is in excess of 90 percent." Obviously it had been explained to him that CellPro's system did not produce a suspension that would be considered "substantially pure" under his initial response.

Among our expert witnesses was Dr. Oliver Press, a professor of medicine at the University of Washington, and the acting director of UW's experimental High Dose Chemotherapy program. Ollie Press was a highly respected oncologist, an aggressive transplanter. Although he was affiliated with the Fred Hutchinson Cancer Research Center, I had never met him.

Ollie Press was uncomfortable spending time in a courtroom. He much preferred to be with his patients. He had never previously testified in a trial, but he agreed to appear as an expert witness in this case because he believed CellPro was getting "a raw deal." As an expert in transplant technology, he knew full well that the Ceprate system owed nothing to Civin or Baxter.

Our lawyers felt strongly that we needed a doctor who was actually treating cancer patients with our device to testify. In the coldness of legal maneuvering, it was easily overlooked that lives were at stake in this case.

There was no advocate for the patients benefiting from our technology. That is by design; juries are not supposed to be swayed by emotional appeals.

Ollie Press was a very impressive witness. He exuded integrity. He spoke clearly and matter-of-factly, and nothing in his testimony even hinted at any personal involvement in this dispute. He was speaking for the people who had no voice in these proceedings. He simply laid out the facts as he believed them to be. As I watched him testify that day, I knew without doubt that had he felt differently, he would have just as directly made that clear.

He testified that Civin's patent claims could not have been supported by data existing at the time those claims were filed. When asked if the patent claims contained sufficient information for him to gather stem cells and perform a transplant—if they "enabled" him—he said that in his opinion they did not. Either he believed completely what he was saying, or he could have been a very successful actor.

He had flown in on the red-eye to testify. As soon as he got off the stand, he was flying right back to Seattle to transplant a patient. After he got off the stand, I caught him out in the hall and thanked him for his support. We spent perhaps two minutes together. If I had thought about it, I probably would have assumed this was the last time we would meet.

I certainly never could have imagined how soon, and in what different roles, we would meet again.

Dr. Civin testified before I did. Curt Civin was a relatively short man with thinning gray hair who spoke in a slightly nasal tone. Actually, I thought he sounded pretty whiny. It was impossible for me to be objective about his appearance on the stand. I had great respect for him as a scientist. He had done important work. But, just like McKelvie and Baxter's attorney Don Ware, I wondered if he really understood the possible consequences of this trial. Basically, his testimony was intended to support the claim that we had stolen his invention to make money. I don't think it was clear to the jury that he was earning a royalty on Baxter's system and had just as much a financial stake in the outcome as me or anyone else.

I thought he tried much too hard to appear to be the wounded party. To me, at least, he seemed overly solicitous to the jury. During his testimony we introduced an unsolicited letter from a transplanter at Johns Hopkins, explaining that he preferred our system to Baxter's Isolex. Asked to respond, Civin got very upset, claiming we'd taken the letter out of context, and lost his cool on the stand. I don't think he helped their case at all.

I was the final witness. By the time my name was called, I felt like the football player standing on the sidelines, pleading, "Please, coach, put me in." I was ready. As I was sworn in and sat down in the witness chair, I was considerably less nervous than I had expected to be. My attorneys had done a fine job preparing me for my testimony. I thought that my direct testimony went well. During cross examination when Don Ware asked me a question he looked me right in the eye—it's possible he was trying to intimidate me—but I quickly discovered that if I stared right back at him, he would look away.

It's natural, I suppose, to dislike the attorney for the other side in a trial, but Don Ware made it easy. He is a small, slender man who wore large glasses and had a Lincoln-like light-colored beard. I know I can't be objective about him, but I thought of him as the whiny little kid who'd spent much of his childhood getting beat up in the playground, then much of his adult life getting even with people because of it. Ware was a Harvard Law School trained litigator, the kind of pit bull who would bring everything to bear to win his case. I remember most his tenacity, his willingness to fight over even the smallest point.

Ware and I had some ugly times during pre-trial depositions. As any good lawyer, he knew the answers he wanted from me and he was relentless in pursuit of those answers. Fortunately, I had been well-prepared by Coe Bloomberg, and I was ready for him. One day, I remember he asked a rather complicated question and instead of responding I said, "I want to speak to my attorney before I answer that." Then Coe and I started to get up to leave the room.

Ware snapped at me, "Don't leave this room until you answer that question."

I glared at him. He was literally quivering with anger. I slowly stood up, never taking my eyes off him, and then Coe and I walked out of the room.

Ware understood that my testimony would not benefit Baxter, so he attempted to provide a reason that I might lie about the facts. Early in the cross-examination he asked me how many stock options had been granted to me. Ahh, money. What better reason would I have to lie to a jury?

"Over the course of my employment," I responded, "I believe the total is 385,000 shares, at a range of prices."

He handed me a blurred copy of a page from CellPro's most recent proxy statement. I didn't recognize the document. When I noted that it was difficult to read the numbers, Ware asked, "Do you have difficulty, Mr. Murdock, in telling us how many stock options you have in CellPro?"

In fact, this document made little sense to me. It seemed to indicate that I actually held options on 532,813 shares. Financial statements can be confusing, but I knew that wasn't accurate. "I am sorry," I replied, "I would have to go back and compute that number."

The message he was trying to send the jury was that I was lying about my financial stake in CellPro, and if that were true, then certainly the rest of my testimony was questionable.

I had been ready and anxious to testify. I believed my testimony would be pivotal. But this line of questioning shook me up. The facts weren't particularly damaging; it made no difference in this case how many stock options I held. But what it did do was cast a shadow over my credibility. If I lied about my own financial interest in the company, it made sense that I would lie to protect it. If Bob Weiss was right, and in the end the jury was simply going to decide who was telling the truth, then I was afraid that Ware had accomplished his primary objective: he'd made me look like a liar. As I was learning, it's difficult for a jury to determine the difference between a confused witness and a liar.

I was not permitted to confer with our attorneys during the lunch break, so I spent my time trying to recalculate the figures, trying to make sense of them. At the same time, Weiss and Kiley also tried to resolve the discrepancy. During my redirect questioning, Weiss just nailed Ware. It was as if Perry Mason were doing the questioning. Holding the same piece of paper in his hand that Ware had used, Weiss asked, "While it's hard to read, in the second column, the one that he [Ware] is referring to, that he indicated was 532,000 shares?"

"I remember that," I replied.

"But this right-hand column, it's not the number of shares . . . it's value?"

Ware had been completely wrong. The number to which he had referred indicated the dollar value of the shares I held, not the number of shares. I almost laughed out loud. But it is not considered proper to laugh at the opposing attorney. "Oh, you're right," I agreed, "I didn't even notice that at the time."

"So Mr. Ware was doing an apples-and-oranges comparison?" To make certain the jury understood exactly what had happened, Weiss took the sheet of paper, walked across the court, and very dramatically handed it to Ware. I just glared at Ware, who was slumped down in his chair.

Later, Ware came over to the counsel table where I was sitting and apologized to me. Then, to his credit, he repeated that apology in front of the jury. "I did want to say that I looked again at the proxy statement and Mr. Weiss was absolutely right. I read a number off the wrong column. I hope Mr. Murdock will accept my apology."

But the damage was done. It was Mr. Ware's credibility that had been shattered.

As the trial progressed, I had found myself watching the jury, trying to detect even the slightest reaction. At one point during my testimony, Ware asked me what I assume he thought was a rough question: "Now, Mr. Murdock . . . would you agree that the market for a stem-cell concentrator would not exist had Dr. Civin not discovered [stem cell] CD-34 antibodies?"

This was the issue over which this battle was being fought. "I don't think that's true at all," I replied, "I think what is enabling here is we finally have, in the area of cell therapy, a system that people can use to do things. And in fact they are using it. I think that is evidenced by the 2,100 patients that have already been treated. That would not have been possible with the My-10 antibody alone that Dr. Civin produced."

Ware pointed out that without the discovery of the stem cell and stem-cell antibodies, we could not have created the Ceprate system, but I thought I'd made our case. In fact, I remember that when I finished my answer I glanced over at the jury and one of the jurors, a young woman sitting in the upper left hand seat, was smiling.

Inside, I smiled too.

The jury deliberated for parts of two days. I spent that time in the courtroom waiting nervously. Jury watch, it's called. I was very apprehensive, at times even pessimistic. They were the longest two days of my life. Over and over we rehashed the testimony. In my mind it always came out the same, we always won a unanimous verdict. During that time for some reason I remembered a scene from *Gone With the Wind* in which the people of Atlanta are standing around waiting for reports about the battle at Gettysburg. I pictured all the people at CellPro standing around in large groups waiting for the phone to ring, or waiting for a story to come across the wire that would tell them their future.

The fact that Kiley spent much of this time drafting two press releases, one to be issued if we won, the other to be handed out if we lost, did little to make me confident. Our lawyers seemed confident, but subdued. I wanted them to tell me that they were sure we'd won, and then I wanted them to convince me they were telling me the truth.

On the afternoon of Friday, August 4, the jury walked back into the courtroom. I was petrified; I couldn't recall ever being more anxious. I'd been told by lawyers that if jurors looked at you as they took their seats in the jury box, you'd won the case. As this jury entered the courtroom, some of them looked right at me; unfortunately, others looked at Baxter's table of lawyers.

As we waited for the verdict to be read, I clasped my hands so tightly in front of me that my knuckles turned white. "All rise," the clerk ordered, and McKelvie entered the courtroom in his black robes. Finally, as I'd seen so often in the movies, the foreman stood and told McKelvie that the jury had reached a verdict. She handed it to the clerk, who began reading it. This verdict wasn't simply for the "plaintiffs" or the "defendants," guilty or innocent, CellPro or Baxter. For some reason Baxter had pushed hard for a highly complicated jury form consisting of 103 different questions. The ballot was a very confusing; this format allowed the jurors to return numerous different combinations of verdicts.

The clerk began reading their verdict form: "One. Do you find that the plaintiffs have shown by a preponderance of the evidence that defendants CellPro's making, use or sale of the 12.8 antibody literally infringes any of the following claims of the 204 patent? Claim one. 'No' is marked. Claim two. 'No' is marked.

"Question two. Do you find that plaintiffs have shown . . ."

The clerk droned on and on. I listened carefully, my fingernails practically digging into the table, but I didn't have the slightest idea what these verdicts meant. I didn't know whether we'd won or lost. I whispered urgently to Coe Bloomberg, "Well? Did we win or lose?"

The smile on Coe's face grew wider as the clerk read each response. Finally he poked me, "Looking good," he said, "really good." For Coe, that was a great burst of optimism. The clerk finished reading the entire verdict and I still hadn't figured out whether we'd won or lost. "We won it all," Bloomberg said. "We got it all."

We'd won a unanimous verdict on all 103 questions. The jurors had heard the evidence and decided that they believed us. The jury found that all four of the Civin patents were invalid because they did not enable others to do the same work, and that even if they had been valid, the Ceprate system did not infringe on their property because our stem-cell suspensions were not "substantially free" of other cells. It was a complete victory.

In the hallway directly outside the courtroom, Tom Kiley had arranged for a Lyon & Lyon paralegal to hold an open phone line with Larry Culver, who was in the office.*

The courtroom remained absolutely silent as Judge McKelvie stiffly went through all the perfunctory business of the court. He thanked the jurors and released them. Just before handing the case to the jury, he had told the lawyers, "Like my brothers and sisters in the Superior Court, I am inclined in jury trials to let the jury have the first shot at an issue, since in most courts that tends to resolve the issue, in most types of law." But oddly, at the end of the trial he told our attorneys, "You need to give those people plenty of time to consider post-trial motions." I was so excited I barely heard him and I certainly didn't understand the deeper meaning of that warning.

The moment the judge left the bench, the courtroom seemed to explode. As one, we all leaped up and started hugging each other. Ben came rushing from the spectator section and threw his arms around me. It was a moment of pure joy. I grabbed Coe Bloomberg, one of the most reserved human beings I've ever known, and he began saying softly, "Rick, I really appreciate the confidence . . ."

I paid absolutely no attention to him. Goliath was vanquished! It was time to party! The sword that had been hanging over our heads since the day the company was founded was gone! The years of frustration, of failed negotiations, were ended. Our biggest problem had been blown to pieces. Now we were going to fight Baxter in the marketplace, and I had

---

* As soon as the entire verdict had been read by the clerk, Kiley went out to a pay phone and told Larry to stop trading in our stock. Standing next to Tom at the other pay phone was a "courtroom observer." She was yelling into the phone "Buy CellPro! I don't care. Just buy as much as you can!" Kiley held the receiver out so that Larry could hear this exchange. Within minutes Larry contacted NASDAQ and trading in CellPro stock was suspended.

absolutely no doubt of that outcome. We were going to be able to do great things.

After hearing the verdict from Kiley, Larry Culver called Josh Green. "The jury just came in with a unanimous verdict in our favor," he reported. And then he broke down in tears of relief. Green let out a loud war whoop—and then admitted he had been completely wrong about this lawsuit. Kiley had been right, Kiley had known it all the time, Kiley had done it again. Amazing, Green thought, absolutely amazing.

Baxter's attorneys were absolutely devastated by the decision. The speed at which the jury had returned the verdict made it obvious that there had been no indecision, no real debate. They had responded to 103 complicated questions in less than two days. After the verdict was announced, the courtroom quickly emptied, but in the spectator section, Curt Civin sat motionless. I couldn't begin to imagine what he was thinking.

As we left the courtroom, like a team that has just won the Super Bowl, I glanced back. Don Ware was still sitting at the plaintiffs' table, gathering his papers. The clerk of the court told one of our attorneys that a half hour after everyone else had left the courtroom Ware was still sitting there looking at his papers.

We returned briefly to the "war room," the office we'd set up in a hotel, to gather our papers. Everyone who had worked on the case was exuberant, everyone except one of our paralegals, a former pro football player who seemed unusually subdued. Noticing that, Ben asked him, "Aren't you a little bit happy that we won?"

"Oh yeah," he said, "I'm very happy."

"Well, you're certainly not showing it."

Then he explained. While playing football, he'd learned that in one game everything might go perfectly, but in the next it might all fall apart. "It was a cycle of incredible highs and devastating lows. Some people can live that way, I didn't want to. I learned to accept the happy medium. Maybe you don't get those real highs that way, but you also don't have to deal with those terrible depressions."

It seemed like an awfully sad way to go through life. At least it did that afternoon.

We partied deep into the night. We went to an Irish pub and drank barrels of Guinness. Even Coe Bloomberg, a man who could hold his drink with the best of them, was slurring his words—but still quietly. "I'm glad this was the first trial you've seen," he told Ben. "This is the way the legal system is supposed to work." All of us were absolutely exhausted. We'd lived this trial day and night for two weeks. For me, the existence of my company had been at stake. But for the attorneys, even though they had become personally involved with CellPro, this was still simply another case. In the end, win or lose, on Monday morning they would begin fighting another case.

Ben and I never got to sleep that night. Early in the morning we flew to Chicago to meet Patty and Jamie, and we all flew to the Caribbean for a sailing vacation. The following Monday, while sitting on the boat, basking in the Caribbean sun, I got on a cell phone and conducted a conference call with all of our stock analysts. It was a spectacular moment. One of the happiest days of my life. Finally, finally, I was able to give positive news to our analysts and shareholders. Asked what the next step would be, I explained that there certainly would be post-trial motions, but then the judge would enter the final verdict in the case, "and then we're going to pursue the antitrust aspect of this case."

When this conference call was done, I turned off the phone, leaned back, and closed my eyes. For the first time in many months I was able to relax. I had been through four battles and survived them all: internal dissension, Corange, the threat of disapproval from the FDA, and the court battle in which Civin's patents had been ruled invalid. For the first time since I had become CEO I could bring my full attention to building the company. Or so I believed.

I didn't get much time to celebrate our victories. Several months later I was standing in front of a mirror, shaving, my mind already far into the day. Suddenly I noticed an enlarged lymph node in my neck. I gently ran my fingers over it, surprised I hadn't noticed it before. "Boy," I thought, "this thing is huge."

This was the lump that somewhere, deep in our souls, we all fear.

PART TWO

THE EXPERIMENT

TOSS A COIN, WATCH IT FLY,

WHO WILL LIVE, WHO WILL DIE.

—An old children's rhyme

5 NO ONE KNOWS WHEN CANCER BEGINS. OR WHY. No one knows when the first cell mutates and divides and eventually slips unnoticed into the bloodstream, there to begin its deadly journey. Maybe I was laughing at the moment it happened, or cheering for Jamie in a swim meet; I might have been caught in a Seattle traffic jam, or sitting in Judge McKelvie's courtroom.

I had plans, great plans, for CellPro. We were among the very few biotech start-ups to remain independent while successfully bringing a product to market. We were the dominant company in a market estimated to eventually be worth a minimum $100 million annually. We were helping save lives, and we expected to save many more. My days, so recently overflowing with legal problems and internal problems and regulatory problems and financial problems, were suddenly filled by expansion plans and product development decisions and marketing discussions. For the first time in so long, I was doing the business of everyday business.

I have heard it said that human beings make plans, and God laughs.

I was concerned about the lump, but I did nothing about it. It'll go away, I figured. It's nothing. An infection. I'll have it checked if it doesn't go away. The matters of the day quickly occupied my attention. It was nothing, I was sure of it. Nothing to worry about, I tried to convince myself. I didn't even mention it to Patty. I was confident it would go away. I refused even to consider the possibility that it might be the first symptom of cancer. It was impossible to believe that the CEO of a company dedicated to treating cancer would actually contract cancer. That was beyond irony.

A week later, while taking a shower, I found the second lump. This one was in my groin. I felt the first twinge of real concern. This time I told Patty about it. As I knew she would, she immediately started worrying. She was insistent that I see a doctor right away. "Richard, you have to get this checked out. You can't wait on something like this."

She was absolutely right, of course, I knew that, but as always I was very busy at CellPro. Every night Patty would ask me if I had made an appointment with the doctor, and every night I promised her I would do it the next day. Finally I couldn't put it off any longer. I was still sure it was nothing, but these lumps were not getting smaller.

I didn't mention the lumps to anyone at CellPro. These were scientists and technicians who worked every day with cancer cells, but never with patients. They were researchers, not doctors. Besides, I didn't want to mix my personal life with my corporate identity. How ironic that decision turned out to be.

Between appointments one particularly busy day, I went to see the only internist I knew in Washington, Dr. Debra Barto. I had gone to see her for a routine physical two years earlier. With the exception of an occasional cold, I hadn't been sick enough to need a doctor since we'd moved to Seattle. I expected the doctor to tell me that it was some sort of infection, but she didn't. "This bothers me," she said. Having previously worked at the Hutch, she had examined thousands of similar lumps. "I think you should have this lump in your neck biopsied. Lumps like this could be the first sign of lymphoma, so please don't wait too long."

I stiffened. Lymphoma was a type of blood cancer that affected white blood cells called lymphocytes. It was a formidable enemy—and I knew enough about it to be immediately alarmed. In my work I'd heard it discussed millions of times—but never in relation to my own life.

She recommended Dr. Linville, a local surgeon with whom she'd worked. Coincidentally, Dr. Linville's son had played on the same baseball team as Ben. We knew each other from the bleachers. After examining me, he did not seem overly concerned. "Swollen lymph nodes are pretty common," he explained, telling me precisely what I wanted to hear. "They're caused by an infection. This one in the groin actually could be related to athlete's foot." He suggested we watch it over the next few weeks, and I made an appointment to return.

Because I had insisted to myself that this was nothing more than an ordinary infection, I wouldn't allow myself the joy of feeling relieved. But I was very relieved. The following week I flew to Brussels for a meeting with our European marketing people. It was the worst trip of my entire career. Owing to bad weather in Europe, it took me almost an entire day and night to get there, and when I finally arrived, my luggage was lost somewhere in England. I had only the damp, wrinkled clothes I had been wearing since early the day before. So it didn't surprise me that I completely lost my voice. I conducted the meeting by whispering to the person sitting next to me, who loudly repeated my words. By the time I got back to Seattle I was terribly sick. I couldn't remember ever being so sick. *I'm physically and mentally exhausted,* I decided. *My mind is finally allowing my body to deal with the incredible stress of the previous year.* With the successful completion of the trial, I had the luxury of time to be sick. That's all it was, I assured myself.

I saw Dr. Linville again just before Christmas. This time Patty went with me. As we drove to his office we carefully avoided talking about the only subject on our minds. As Dr. Linville gently explored the lump with his fingertips, I tried without success to read the expression on his face. Finally he said, "It looks like that lymph node in your neck has gone down a little bit." I was impressed by his ability to calibrate its size simply by touch. In an office replete with the most sophisticated technology available, there was something strangely comforting about his confidence in his own observation, experience, and intelligence. "I really don't think this is anything." He suggested I spend a restful Christmas with my family, and return for another examination after the first of the year.

In a diagnostic situation like this, a few weeks make no difference in long-term outcome. But they certainly do to the patient. I was caught between wanting to know immediately and never wanting to know. After leaving Linville's office, Patty and I drove directly to the public library. In the reference section we found a book that described the symptoms of most major diseases. We leaned over a table, shoulder to shoulder, learning about lymphoma. It outlined the characteristics of the disease, explaining its causes and how it progresses, then gave the terrifying survival statistics. While the information was not particularly current, it was still pretty scary stuff.

With the exception of the lumps, my symptoms were sort of vague. My lumps did not precisely fit the description given in the book, but they were too close to be dismissed easily. The one piece of information we read that made me feel a little better was the statement that once the lumps appeared, they never got smaller. According to Dr. Linville, the lump in my neck had gone down.

Throughout the Christmas holidays I tried to push my fear out of my mind, but that was no more possible than trying to ignore a single pebble caught in my shoe. Is it or isn't it? That question was always there. I pushed it into a dark corner and enjoyed a wonderful holiday with my family. Patty cooked for twenty-five people, family and friends. It was a celebratory blowout. CellPro had survived every challenge. We had created a new and vitally important weapon for transplanters, and now we were poised to market it.

On the third of January, Dr. Linville gently touched the lump in my neck. He frowned. "It's bigger again," he said without emotion. "I think we ought to take it out."

Each step in this diagnostic process was very small, but it was beginning to look as though they formed a path. My mind, which is generally organized, was in shambles. I had convinced myself it had to be nothing because I'd led such a healthy life. I had exceptional bloodlines. I'd never had a serious illness. Both my grandparents had lived into their nineties. My father walks two to three miles a day; my mother is a strong, healthy woman. I took care of myself, I exercised regularly, I watched my weight, I tried to eat the proper foods. It just couldn't be cancer. Not me.

Linville gave me a local anesthetic and removed the entire lymph node. It wasn't easy. He struggled with both hands to pull it free. I could feel him pulling, but I felt no pain at all. Instead I felt very disconnected, as if I were watching this happen to someone else. "There's some pus coming out," he said. "That's a good sign that it's just an infection."

When he finally managed to get it out of my body, I didn't even look at it. After stitching up the small slit in my neck, he said, "Just wait here. I'm going to take this down to the lab. They'll send it out for analysis. I'll be right back."

But he didn't come right back. He walked down the corridor to the pathologist and did not return. I waited and waited. And I started getting very nervous. Finally his nurse told me, "Dr. Linville said he'd have the results in two days, and he'd like you to come back then."

The incision was covered with a large white patch. It stood out against my skin like a bright flag against a dark sky. It was impossible not to notice it. Naturally, everyone at the office was deeply concerned. The first person I saw was Joe Tarnowski. Joe made a big show of carefully examining the patch, then set the tone for everyone else by wondering aloud, "Gee, Rick, if you look like that, I wonder what the other guy looks like." That was typical of the heartfelt sympathy I received from my colleagues.

"It's nothing," I told him, "I just had a little thing removed." There was nothing anyone at CellPro could do to help, and I didn't want them to worry unnecessarily.

Later, Joe admitted that he had been very concerned. From the beginning the entire scenario had sounded ominous to him.

A few days later I stopped by Linville's office on my way to work. I wanted to get this meeting out of the way as fast as I could. The boys were at home and Patty had things to do for them, so she did not come with me. Both of us were trying to make life as normal as possible. I waited in an examining room for Dr. Linville to walk in with the results of the biopsy. Linville's a big man, and I expected him to burst into his office, casually toss a folder on the table, and tell me, "There's nothing there. Nothing to worry about. Go about your business." We'd shake hands and maybe laugh a little too hard over something amusing, and the next time we'd see each other would be in the bleachers.

But that's not what happened. As he walked into the room, I stood up. I was about to have my sentence pronounced. He kind of put his hand on my shoulder and said calmly, "Look, there was some lymphoma there. I want to take you down to the oncology people here. We're not sure what it all means, and there're some other things we'd like to do. But we did find some lymphoma there."

I was stunned. Lymphoma? I was in total shock, tempered only by denial. I followed Linville out of his office, down a staircase, to the other side of the building, where the oncology group was located. We walked from one side of the building to the other. It was the perfect metaphor: I had crossed the line that separates those people who have never battled cancer from those people who have faced this terror. Whatever happened from that moment on, I could never return to the other side. The war for my life had started.

I fought to contain an overwhelming sense of panic. The word *cancer* had long been an important part of my daily vocabulary. I had used it many times every day. I understood the science of cancer quite well. I could speak intelligently about T-cells and B-cells and lymphoma and leukemia and breast cancer. I could describe in good detail various treatment protocols. I could quote cure rates and remission rates. At conferences I could sit and listen to scientists discuss complicated cancer experiments and understand every word they said.

But it was as if I had never heard that word before. In an instant the word *cancer* had taken on an entirely new meaning. For the first time that word carried with it an ominous personal threat.

Despite my knowledge of the disease, I actually had spent very little time with cancer patients. I remembered once walking through the bone-marrow transplant ward at Johns Hopkins and looking at the patients with great sympathy, but with even greater relief that it was not me sitting there. And I remembered being at the University of Colorado Cancer Center to help design our Phase III clinical trials. Our chief investigator there, Dr. E. J. Sphall, wanted us to see a breast-cancer patient getting back her purified stem cells. Elizabeth Sphall was the first doctor to use the Ceprate system on patients, and conducted most of our Phase I and II trials.

Because the patient's immune system had been destroyed in preparation for her transplant, she was living in a clean room, a germ-free environment. We all put on masks and gowns, then followed Dr. Sphall into the room. The patient was sitting up in her bed. She had a Hickman catheter, a permanent needle port through which she received fluids, implanted in her chest. And as a consequence of chemotherapy, all her hair had fallen out. But this woman did not look terrible, particularly for someone who had failed several previous therapy regimens and whose only hope for life was a successful transplant. But she did not look normal, either. Cancer patients have a very distinct look, particularly those who have endured chemotherapy. We stood far away from her as Dr. Sphall put a syringe containing a collection of her purified stem cells into the catheter and pushed down the plunger, forcing her cells back into her body. Interesting, I thought, that's very interesting. I was far more concerned about the effectiveness of the treatment than the survival of the patient. I don't think I even knew her name. I certainly never learned if she had survived. I didn't want to know.

Now I was to become the person in the bed as others stood far away from me and watched dispassionately. I sat down across a neat desk from an oncologist and he began discussing the side effects of chemotherapy. As I listened to him, I wondered, *What am I doing here, listening to this guy?* With all my scientific understanding of cancer, I didn't have the slightest idea what to do. My education had not covered becoming a patient. He said, "We have a meeting once a week where all the new lymphoma cases are presented to a committee at the University of Washington chaired by a doctor named Ollie Press. What I'd like to do . . ."

I stopped listening to him. Ollie Press? I knew Ollie Press. It seemed like only days earlier when we'd been standing in the hallway outside the courtroom. ". . . one of the most respected oncologists in the world," the doctor continued. Ollie Press had testified as an expert witness in the trial. He'd been very supportive of our case, just great, wonderfully knowledgeable and sincere.

I decided instantly I wanted Dr. Press to treat me. I believe the oncologist sitting with me was disappointed because he wanted to treat me himself.

I reached Ollie Press in his office at the University of Washington Hospital about an hour later. I was pretty blunt. "Ollie," I said, after reminding him who I was, "I was just diagnosed with lymphoma."

"Oh, Rick," he replied, "I'm really sorry to hear that." I explained the situation and told him I'd like very much to be treated by him. He was sort of noncommittal, but made an appointment to see me.

Finally, it was time to tell Patty. That was tough. But I knew that once I told her I would feel a little better. Somehow she would find a way to reassure me, to assuage the panic I could feel rising inside me. I called her at home, where she was in the kitchen preparing lunch for several guests. I don't remember my precise words, but I told her that I'd seen Linville and the results . . . the results . . . I do remember that my voice broke as I told her. ". . . the results were positive. It's lymphoma."

She was very quiet. Patty first listens, then feels, and only later responds. When she finally responded, she told me exactly what I needed to hear: This was simply another problem that we would deal with and get through. We'd talk about it more when I got home. There was never even a wisp of despair or fear in her voice. It was a brief conversation, and then I went on with the business of my day.

My son Jamie was leaving to go back to school at Worcester Polytech the next day so we decided to tell the boys that night. I had been very low-key about the biopsy—so low-key, in fact, that Jamie had completely forgotten about it. "We had a bit of bad news today," I began. "Remember that biopsy I told you about? Turns out it was positive." Jamie told me later that he didn't quite understand what I was telling him, but that he knew enough about cancer to know this wasn't a good thing.

Ben asked, "How serious is this?"

I've never lied to my children. "Very serious." Life-threatening, I told them, but assured them that I intended to be around for a long time. Which was exactly the way I felt.

The boys eventually went upstairs, Jamie to pack and Ben to study for a physics test. They spent some time discussing it. Jamie wondered aloud if he should stay home, but we assured him that there was nothing he could do to help me. The next morning he left to return to college.

Telling people you have cancer is not a particularly easy thing to do. What do you say? "I don't want to bother you, but I may be dying"? The last thing I needed or wanted was sympathy. If I hadn't had to tell the people with whom I worked, I probably would have kept it secret until that was no longer possible. But as the CEO of CellPro, I had a responsibility to my board of directors and my employees. I don't remember the order in which I told people. That's all sort of one great blur. But I do remember that Larry Culver was the first person I told. In addition to being extremely supportive, Larry was my friend. "God," he said, closing his eyes. "God, I was afraid of that."

As usual, Joe Lacob was in the middle of at least twenty projects when I reached him at Kleiner Perkins. "You're going to have to take a minute," I said. "I want you to close your door and sit down."

I love Joe Lacob in a lot of ways. Sometimes I think of him as my crazy brother. I know him well—so I know that when he thinks he's about to hear bad news, he gets very nervous. "Uh-oh," he said. "What are you going to tell me?"

"This is going to be hard to talk about," I continued, "but I've just been diagnosed with lymphoma." Joe was incredulous; he admits that he had to fight not to cry. I was, he had often told me, the healthiest person he knew. Joe and I often used to kid about sports. He would describe me to people as "a guy who actually participated in triathlons." That wasn't true, and he knew it. It was actually his way of describing my determination; if there was something I wanted to do, somehow it got done.

Joe remembers that my voice broke as I told him, and perhaps that was true. In only a few years we had become close friends, close enough that I would show him my vulnerability. But when I told most other people, I remained stoic. In my words and actions I expressed great determination and showed very little emotion.

Several days later I sent an e-mail to all our employees. It was a simple message: I have been diagnosed with non-Hodgkins lymphoma and will begin chemotherapy very shortly. Undoubtedly, from time to time I will miss some days, but my prognosis is very good. My schedule is going to change a bit, but otherwise let's not let this deter us from the things we need to be working on. Let's just keep going.

Among the people who received that message was a thirty-six-year-old scientist named Nicole Provost. Several months earlier, Nicole had finished development work on a model system for isolating stem cells in peripheral blood, and had begun working on our tumor-cell-purging project. This was not a priority project. Like just about everyone else at Cell-Pro, when she received my e-mail she was surprised and saddened. After the long legal battle, after surviving business problems, she wondered, what else could possibly happen to this company? But she never imagined that within months that e-mail would so drastically alter her life.

CellPro's employees sent me numerous e-mails, cards, and notes of encouragement. Few people ever have had a rooting section as supportive as mine. The question then became, Do we go further and inform the stockholders? That was a tough one to answer. What is the obligation of a company to its investors? That news would probably have negatively affected the stock to some degree. After debating it, the board decided not to announce it publicly unless and until it affected my ability to run the company. Several years later, in a *Wall Street Journal* story concerning CEOs who became seriously ill, we were severely criticized. But I've always believed it was the proper decision.

The day after being diagnosed, or perhaps it was two days later, Patty and I drove downtown to the University of Washington to meet with Ollie Press. I had been to the UW cancer center many times, and I'd always just breezed through the lobby on my way to meet a doctor to discuss the Ceprate system. For the first time I had to stop at the desk for instructions. I had no idea how to be a patient. I didn't know where to register, where to go; I didn't know anything about filling out insurance forms, or making appointments for tests. For the first time I had to sit in the waiting room.

The last time I'd seen Ollie Press had been outside a federal courtroom in Wilmington, Delaware. On that occasion he'd been dressed in a suit and tie. This time he was sitting comfortably behind his own desk, wearing a white lab coat. As I was soon to realize, I had entered my own Land of Oz. There I would meet many people I already knew, but each of them would be cast in a new role in my life. The expert witness Ollie Press was now my doctor. He was to be my guide through this world. Months earlier he'd helped save my company, but now . . .

As I sat in his office with Patty, discussing my fate, I felt completely disconnected. It was so difficult to really comprehend that we were talking about me. About me! Ollie began by asking a series of questions concerning my family history and general health. He was just a bit surprised that I had no history of cancer in my family, as lymphoma is a genetic disease that tends to recur in families. Then he gave me a complete physical. Except for the fact that I had a potentially fatal disease, I was in excellent health.

Ollie inspired confidence with his honesty. He classified my disease as an intermediate-grade, cleaved-B-cell lymphoma. There are about twenty different types of lymphoma, he explained, and "if you're going to have lymphoma, this is one of the best types to have. It responds to therapy."

By this time I'd gotten on the Internet to learn as much as possible about the way my own body was trying to kill me. The concept that a deadly disease was growing, minute by minute, inside my body was terrifyingly difficult to accept. I looked and felt perfectly healthy. My primary symptoms were two enlarged lymph nodes. It hardly seemed enough to kill me.

Oh, but it was. There was an extraordinary amount of information to be found on the Web, information I would not have been able to gather only a few years earlier. I read abstracts from recent papers, copies of articles from journals, the texts of speeches delivered at recent seminars. I learned quite a bit about this enemy I was battling.

Lymphoma is a deadly disease. In a healthy body, when a cell divides, each new cell is an exact duplicate of the parent cell. Each human cell contains twenty-three chromosomes in a very specific order. Every chromosome—these are the carriers of genetic information located in the cell nucleus—is in exactly the same place in every cell. Lymphoma is caused by a translocation of chromosomes in a cell, a confusion during cell division that causes parts of chromosomes to end up in the wrong places. They're out of order. It's sort of like a copy machine suddenly going out of control and transposing two different paragraphs to change completely the meaning of the text. Specifically, the #11 chromosome and the #14 chromosome have broken and rejoined abnormally, so that part of #11 is where #14 should be and part of #14 is where #11 should be. No one knows why this happens, why the chromosomes break; there are all kinds of theories ranging from genetic to environmental to just simply bad luck, but this translocation screws up the mechanism that regulates normal cell reproduction. Instead of cells reproducing at their normal rate, the mechanism that controls cell growth is turned off, so lymphatic cells, the soldiers of the immune system, reproduce much faster than healthy cells, they grow out of control, and they live eternally.

In lymphoma they grow so fast they overwhelm the body—the visible lumps are simply collections of cells that the body can't process—and eventually clog up the organs, preventing them from functioning normally. Think of a copy machine that can't be shut off, printing thousands of copies, tens of thousands of copies. The immune system is compromised. These cells invade the vital organs. Patients die from infections the body is unable to fight, or from the inability of organs like the kidney and liver to perform their normal function.

The only way to fight lymphoma is to stop these cells from multiplying by killing them or eliminating them from your body. Prior to chemotherapy, there was no effective treatment; the only thing doctors could do was make their patients as comfortable as possible until the disease killed them. It was a slow, agonizing path to death.

As I discovered, there is still no single recommended treatment, or protocol, for lymphoma. No two lymphoma patients are alike. They may have the same disease, but everything else about them will be different. An oncologist is like a medical tailor; after taking the measurement of the patient's disease, he designs a treatment regimen to fit his particular case. But all lymphoma treatments have the same objective: eliminate the tumor cells. Kill them with chemotherapy, radiation, or immune therapies in which designer antibodies are created to search out specific tumor cells and destroy them. In any way possible, in every way possible, eliminate them, kill them.

My lymphoma was initially diagnosed by Ollie Press as an intermediate-grade, non-Hodgkins lymphoma. Basically, that meant I had an aggressive cancer spreading throughout my body. The long-term prognosis for curing intermediate-grade lymphoma with standard therapy was almost 50 percent after five years, and if that failed there was another 50-percent chance of curing the disease with a transplant. That meant I had a 75-percent chance of living to see my boys grow up.

I never allowed myself a moment's doubt that I was going to be cured. Doubt was a luxury I just couldn't afford. But admittedly there were moments when my confidence was badly shaken. One afternoon as Patty and I sat waiting for my appointment with Ollie, I noticed someone wheeling a distinguished looking man in a wheelchair. The man in the wheelchair was obviously very, very ill. He was barely responsive. Clearly he was going to die very soon. Eventually he was taken into the center.

About a half hour later I was taken into a treatment room. I had to wait there quite some time for Ollie. That was unusual; he was always very prompt. When he finally came in, he said, "I apologize for being late, but one of my patients really has taken a terrible turn for the worse."

I knew he was talking about the man in the wheelchair. It was a sobering thought. This is what happens to Ollie's patients, I thought, he treats them for a while and then . . . I couldn't get the picture of that man out of my mind. While I remained optimistic about my own prospects, here was the reality: People who came into the center *died*. It was just so difficult to believe this was happening to me. I wondered if all patients felt as confident about being cured as I did.

So I wanted those damaged cells out of my body right away. I wanted the vaunted miracles of modern medicine to sweep them out of my body so I could have my own life back.

If chemo failed, patients had no other treatment options until the late 1970s, when Dr. Fred Applebaum, then at the National Cancer Center of the National Institutes of Health, began transplanting patients who had failed chemo. Bone-marrow transplants allowed patients to be bombarded with much larger doses of chemicals and radiation than was possible during normal chemotherapy—doses that killed every blood cell in their bodies. After the chemicals had done their job, the patient's own bone marrow was put back into his body, where it produced new cells. Transplants had become relatively common for lymphoma patients who had failed standard treatment.

I wasn't particularly interested in learning more about transplants. At that point, Ollie saw no need for a transplant in my case. With the type of lymphoma I had, bone-marrow transplants had not proved to be any more effective than conventional chemotherapy. The potential benefits of a transplant simply weren't worth the significantly increased risk.

After reading as much information as possible about my disease, I decided to be treated with a combination of four chemotherapy drugs, known collectively as CHOP, which kill cells at different stages of their growth cycle. But, in addition to CHOP, Ollie Press was conducting experimental trials of high-dose chemotherapy protocols in which he used stronger chemicals; like most aggressive oncologists, he was continually searching for new weapons in his battles with cancer. "I can't tell you they're more effective," he admitted. "There are no statistics to prove that. But people think they're going to result in higher cure rates."

Cancer is a game of statistics. You go with the odds, even while you're convinced they don't apply to you. High-dose chemotherapy made sense. It killed more cells than did standard CHOP—therefore it killed more tumor cells. The downside was that it destroyed blood cells so effectively that I would have to be hospitalized during the treatment, and I would need blood transfusions to replace the cells that were killed. I would definitely get sick, and I would probably get serious infections. "It's going to be a lot harder for you to work, and you're going to have more down time than if you went with standard CHOP," Ollie continued. "It's definitely going to affect your ability to carry on a normal lifestyle."

It was absolutely essential to me that I be able to go to work every day as much for my personal benefit as for the good of the company. The Ceprate device remained caught in the Food and Drug Administration bureaucracy. We'd responded to the FDA's concerns, but still had not received approval. I believed they would be holding hearings within the next few months, for which extensive preparations would be necessary. But primarily because there was no evidence that high-dose chemotherapy was any more effective in fighting my disease than the normal regimen, I elected to be treated with standard CHOP. I would receive at least six doses of very strong drugs over the next several months. Ollie was optimistic that my disease would respond to that treatment.

Until the day I began chemotherapy, I don't think I truly accepted that I had a terrible disease. I was never really sick. I never had any significant clinical symptoms. Except for making time for doctors' appointments, it hadn't affected my life. But within hours of receiving that first dose of drugs, I knew this was for real.

The process takes about two hours. It isn't the slightest bit painful. I was seated in a comfortable reclining chair. A flexible catheter was inserted in a vein in the back of my hand. The nurse checked several times to make sure the catheter was inserted deeply into the vein, because if these toxic chemicals leaked onto my skin they would eat right into the tissue. Finally the catheter was connected to the IV. The drugs were in plastic medical bags, possibly made by Baxter—that was one of their most successful product lines. The bag of chemicals had been mixed in the pharmacy just prior to infusion. The first drug I was given was an anti-nausea drug. Then I was given a very mild sedative—it made me feel very calm, very relaxed. And then I got the chemo.

Patty was with me during each infusion. While these deadly chemicals dripped into my body, we sat and talked as casually as if we were having a nice lunch on the sailboat. When the infusion was done, I felt just a little bit queasy, but I could have returned to work.

In fact, I felt fine until I woke up the following morning. And then I felt bad—very, very bad. I took powerful anti-nausea drugs, but even they didn't completely eliminate the queasiness. The first two or three days after each treatment were tough. I'd be at work and waves of nausea would sweep over me. The only thing I could relate it to was seasickness, and I almost never got seasick. At times I would try to visualize what was happening in my body. It was difficult to think of my body as a battlefield, but that's what it had become. I've seen life under a microscope, I know what cells look like. In my mind I would imagine these chemicals hunting down cells, disrupting their reproductive cycles, destroying them.

For me, far more distressing than the nausea was the loss of my hair. When I began chemo, my hope was that I would lose a few pounds. I figured that was the only positive thing that possibly could come out of this—besides saving my life. I'm about six feet tall and weigh slightly more than 175 pounds. During chemo I lost almost no weight. That was disappointing. What I did lose was my hair, and that was very difficult for me.

I don't consider myself vain, but admittedly I do take great pride in my appearance. I've always related my appearance to my thoughts, I'm an orderly person in my dress and grooming as well as my thinking; one of the passions of my life has been fine clothing. My side of the closet is almost as jammed as Patty's side. I love the feel of a well-made Italian suit. I love wearing a sparkling white shirt with a starched collar. I probably own fifteen suits, all of them either charcoal gray or navy blue, because those are basically the clothes of business, and fifty different ties. Image does count in business.

When I meet someone, I do notice the cut of his clothing, and within a few seconds I know just a little more about this person.

Admittedly, people have made fun of my dress habits—people in addition to my very own sons. At HemaScience, someone once asked me if I kept an ironing board in my office. At CellPro, we instituted "dress-down Fridays," causing someone to remark that, for me, dressing down meant taking off my jacket.

My hair has always been an important part of the image I've tried to project. I've always had a full head of neatly trimmed dark hair—and a handlebar mustache. I loved my mustache. I'd first grown it, and a goatee, when I was at Berkeley, to complement my shoulder-length hair. I loved the fun of it. The long hair was trimmed to business length many years ago, but the mustache just grew and grew, until it became a long, thin handlebar. I could twist it and play with it. I've had it my entire adult life. Patty, my kids, my friends, my co-workers, no one had ever seen me without my mustache. I always sort of thought that the combination of trimmed and combed hair and the handlebar mustache projected the image of someone with the respectability of an established company but the whimsy of a start-up.

I had been warned that my hair would fall out when I began chemotherapy, but the reality of it hit me hard. Unlike Samson, I wasn't losing my strength—instead, I felt like I was losing control. This was a constant reminder that I was sick. The hair storm, as it has been called, arrived unexpectedly one morning. After my first round of chemo, I was scheduled to fly to New York to address analysts at Merrill Lynch. In the shower that morning I started to rub shampoo into my hair—and suddenly it began coming out in handfuls. I broke out in a cold sweat, while still in that shower. Part of my self-confidence disappeared down the drain with my hair.

In retrospect it seems so silly. But that morning I was devastated.

I dried my hair and arranged it to look as full as possibie. I made it through that meeting, but after that I really began to lose my hair. I'd never worn hats in my life, but I decided it was appropriate. Patty and I searched all over for something stylish. I ended up buying four or five different hats, among them a nice fedora and a Panama.

I began wearing a hat almost all the time. Just about the first thing I did in the morning was put on a hat. I wore it in the office. I even wore it on airplanes. Eventually I decided I needed a hairpiece. I always referred to it as a hairpiece rather than what it was, a wig. Many other people shave their heads completely when they start losing their hair, but I just couldn't do it. I wanted hair. I just couldn't give it up. The UW Cancer Center recommended a company in Seattle that made very realistic toupees. I had several fittings, and it took almost a month to make. But when I put on that hairpiece, it was as if I were putting on my confidence.

Also, in the mornings Patty would fill in my eyebrows with an eyebrow pencil.

The temporary loss of my hair was a small price to pay, and when you're going through something like this, it's easier to focus on vanity than survival. I didn't let many people see me completely bald. Jamie came home from school for a few days, and late one night he tripped over a laundry basket and made a lot of noise. Without pausing to put on my hair, I went downstairs to see what was going on. When I stuck my bald head around the corner, Jamie was shocked. He couldn't stop staring at my head. Finally he smiled and said, "You know, Dad, you look pretty good like that. If you bulk up a little, you could probably get a job as a bouncer somewhere."

Perhaps because I made jokes about my hairpiece, other people felt free to do it too. When I started wearing the toupee to work, for example, people began quoting the famous advertising line from the hair-replacement company Hair Club for Men: "I'm not just the president, I'm a client."

After going through three cycles of CHOP, it became obvious these chemical killers were not doing an efficient job. As Ollie explained, "We're just not getting there fast enough." My prognosis was, at best, fair—but there was still plenty of hope. But that all changed at 2:28 P.M. on March 21. Ironically, it was my birthday.

My death sentence arrived by e-mail. "I received a path report and phone call from Sandra Horning [CellPro founder Rich Miller's wife, and a highly respected oncologist]," Ollie Press wrote, "indicating that the Stanford pathologists have revised your diagnosis from diffuse, mixed small cleaved and large cell lymphoma to mantle cell lymphoma.... CHOP is still the most commonly used regimen for mantle cell lymphoma, but this is generally believed to be a much less curable histology unfortunately than DML. We can discuss ramifications...."

Much less curable . . . I sat at my desk reading this message again and again, as if the meaning might somehow change if I read it one more time. All I knew about mantle cell was that it had recently been identified as an unusually difficult form of lymphoma to cure. These cells were especially aggressive and resistant to traditional chemotherapy drugs. After making the diagnosis, Ollie had to stage my disease. The range of staging is 1 through 4, the prognosis getting progressively worse with each stage. Doctors literally pound a long, hollow needle into the soft inner cavity of my hip bone, then sucked out a representative sample of my bone marrow. It's impossible to accurately describe that feeling. There is some pain, but when they aspirated, or withdrew, the marrow, I could feel it in every part of my body, right down to my toes.

Several days later, Ollie informed me that my bone marrow was positive. Tumor cells had invaded the very marrow of my body. My revised diagnosis was stage-4 mantle-cell lymphoma, meaning the cancer had spread throughout my body and into my bone marrow—the blood-producing organ of my body.

With that diagnosis, everything changed. Because mantle cell lymphoma had only recently been identified, few oncologists had any experience with it. Almost without exception, everyone who had fought this disease had lost. The average survival period of patients was less than three years after diagnosis. For the first time I had to consider the possibility I might not survive. It just didn't seem real. I was the guy whose company was making a device to fight cancer. I wasn't supposed to be a cancer patient. That just wasn't possible.

Well, I wasn't going to lie down and die. Or feel sorry for myself. That has never been my nature. My first decision was that I was going to fight every damn malignant B-cell in my body. I was shocked, angry, confused, maybe even terrified, but I knew I had available to me a new technology to which no one else had ever had access. I'd bet my company on this technology, and now my life was going to depend on it.

Ollie told me that he had recently attended a lymphoma conference in Boston at which a group of German oncologists from Heidelberg had presented minimal data indicating that early-stage mantle cell patients benefited from a transplant. They had only transplanted seven or eight patients, but, compared to transplants on late-stage lymphoma patients, the data looked very good.

In the situation I was in, the things I needed to hear most were a message of hope and a commitment to fight. Ollie was giving me both. After I'd completed four rounds of CHOP, Ollie did another complete workup. I waited for the results, but in my heart I already knew the answer. "We've made some progress," Ollie reported, "but not enough. We're just not gonna get there this way. I spoke to Sandra and we both agree that it's time to start thinking about a stem-cell transplant."

The complication, he continued, was that I had about 25-percent involvement of tumor cells in my bone marrow—it was just loaded with tumor cells, as well as tumor cells circulating in my peripheral blood. There was no source of stem cells that had not been compromised. "If this is going to work, we've got to have a good purging technique," Ollie explained, then added matter-of-factly, "I don't have one here."

As he said that, I leaned back and took a deep breath. My only hope for survival was a transplant. Less than a cupful of white blood cells would be removed from my body. Then my body would be bombarded with radiation and chemicals that would kill just about every remaining blood cell—including the malignant B-cells. But that treatment would also destroy the white blood cells that made up my immune system, leaving me susceptible to every kind of infection. Generally, people can survive at most two to three weeks without a functioning immune system before dying from an infection or internal bleeding. But in that cupful of donated blood were millions of stem cells, the Adams and Eves of cells, the cells from which grow all the various types of blood cells found in the body. When these stem cells were returned to my body, they would take root in my bone marrow and begin pumping out the billions of various cells needed to rebuild my blood and immune systems.

The problem was that among those cells returned to my body would be malignant B-cells. Cancer cells, the hidden enemy. Once reintroduced to my body, they could begin to grow. The device we were building at CellPro, our Ceprate system, picked stem cells out of the blood and collected them, leaving behind most of the tumor cells. We could successfully eliminate two or three logs of tumor cells—a log is 90 percent, so three logs is 99.9 percent—but the procedure would still leave many millions of tumor cells to be put back in my body. At CellPro we had also been working on a negative-selection column, a second step in the purification process, which removed the remaining tumor cells from the collection of stem cells. Not a priority project, it had been relegated to our corporate hope chest. I had supported Ron Berenson's decision to put it on the way back burner and concentrate on the more-marketable breast-cancer tumor cell project. It had been a dispassionate decision, and it made good business sense. We gave absolutely no thought to the lymphoma patients who might need this device to save their lives. More patients needed the breast-cancer depletion device and we thought that it would be simpler to get FDA approval. I certainly never imagined that one of the people deprived of the device when he needed it would be me.

In fact, after my original diagnosis I used to kid with Nicole Provost, the senior scientist in charge of the B-cell depletion project, about the possibility that I might need it. At our monthly meeting with the project teams I'd smile and ask her, "So how's the lymphoma purge going? You'd better hurry up, you know. I just might need it one of these days."

Nicole would smile wanly at my little joke. "I hope not, Rick," she'd replied. It was all very pleasant. Not gallows humor exactly, more like test-tube humor.

Only one person in the lab, an excellent researcher named Sharon Adams, was working on the B-cell depletion project and she was only doing it part-time. The B-cell purging device was strictly experimental. It had never been used on a patient, and we had no treatment protocol or FDA permission to use it. We'd used it strictly for laboratory research, and the results had not been particularly promising.

Personally, I couldn't think of a better time to push forward on the B-cell depletion column. "We can use the Ceprate system," I told Ollie, trying to infuse my voice with confidence. "We've also been working on a purging column for a long time. I may be able to use that."

As soon as I got back to my office, I called Joe Tarnowski and asked him to stop by my office for a chat. Joe had come to work at CellPro a little less than a year after me. He'd become a trusted and loyal friend. Joe is a nuts-and-bolts guy, a biological mechanic. He knows how to get things done. If my plan was to work, Joe would have to play a significant role.

Almost from the moment I was diagnosed with lymphoma, I'd kept the people at CellPro informed of my progress. "Well," I said to Joe, about as calmly as I could, "remember the biopsy I told you about? It seems the results are in and they changed the diagnosis to mantle cell lymphoma."

Joe told me later that his heart just sank when he heard those words. He'd attended enough oncology meetings to know exactly what that meant. He knew that my prognosis was extremely poor. But his expression didn't reveal his emotions. "Are you sure?" he asked.

I nodded. "Yeah, the morphology and the markers are there." From the tone of this discussion we might just as easily have been discussing a baseball game—Tarnowski is a big baseball fan, and for several years I shared his season tickets to Mariners games—as laying a plan to save my life.

"Gee, Rick . . ." he started, then stopped. "You know, if you're gonna get this disease, there isn't a better place to be sitting in the whole world than in your chair."

"Ollie's just been back from a meeting where some data were presented, and it turns out the best way to treat it might be a transplant. My strategy is to get a transplant using stem-cell selection—and I want to use the B-cell purging device."

Joe hesitated, then said, "The purging device isn't ready, Rick. It's a long way from being ready for prime time."

I'd spent just about every working minute of the last four years of my life developing and marketing our Ceprate system. I knew it from the molecular level to the printing on the shipping box. I believed in it completely. "I don't care," I said. "I want to purge, and I want to use our system. We'll just have to do the best we can."

"What's the time line?" Joe asked.

"I need it by May. June at the latest." That was only three months away.

I think Joe smiled at that. All I was asking him to do was accomplish the impossible. We'd spent several years working on the B-cell purging system without getting it to work. Now I was telling him I needed it in three months.

Most science is dispassionate. At CellPro we rarely had any direct contact with the human beings who might benefit—or die—from the success or failure of our work. In the development of the Ceprate system, patients were simply the end product, statistics we needed to move forward with our clinical trials or to get final FDA approval to go to market. None of those people had a name or a face, we didn't know anything about their families or their hopes and dreams. The doctors conducting our clinical trials sent us reports—this many patients showed improvement, that many patients failed to respond to treatment. It's easier to go home at night when you don't know the patients' names.

I was changing the rules. For the first time in the professional lives of most of the employees of CellPro, we were going to have a front-row seat in a clinical test. And the patient was not going to be a number, it was going to be the boss.

"I've got to do a little homework to see where we are on this thing," Joe continued finally. "But you know that at best it's gonna be held together with spit. It's not gonna be perfect." In his mind, Joe had already accepted the challenge and was making plans to meet it. "We haven't even selected the antibodies." He paused again. "This is really going to be experimental."

The irony of the situation was astounding. Beyond astounding. Cell-Pro, my company, was the only place in the world that had the technology that might help save my life. Until this moment I had staked my professional career on CellPro's technology, now my life was going to depend on it. I would be putting my faith in my business in a way few people before me had ever done.

I was going to be the guinea pig. In the language of clinical tests, I would be patient number one.

6 I WAS GOING TO TAKE THE NEXT STEP ON A PATH
that was already thousands of years old.

Mankind has always wondered at the powers of blood. The ancient Greeks considered blood one of the four "humors" that controlled health and disease. In ancient Rome, aspiring warriors would drink the blood of slain gladiators and bulls in the belief that it carried with it the courage of the dead. A thousand years ago, the Anglo-Saxons knew enough about the healing power of blood to use leeches for bloodletting. But it wasn't until 1665 that blood was successfully transfused between two dogs.

In 1667, in an experiment conducted under the auspices of the Royal Society in London—acknowledged as the birthplace of modern science—approximately nine ounces of blood were successfully transfused from a sheep into a fifteen-year-old boy. "The patient," it was recorded, "found himself very well upon it, his pulse better than before, and so his appetite. His sleep good, his body as soluble as usual, it being observed that the same day of his operation he had three or four stools, as he used to have before." This was the first human blood transfusion in English history.

A year later, French physicians in Paris found that transfusions between animals and humans did not work, and caused such attempts to be outlawed.

The first blood transfusion in America was reportedly done in 1795 in Philadelphia. While experimental transfusions conducted by English doctors in the 1800s sometimes proved beneficial, no one yet understood why blood transfusions worked splendidly with some patients but proved fatal in others. As late as 1880, doctors in the United States were transfusing milk into human beings as a blood substitute.

The history of successful blood transfusions begins in 1901 with the discovery of the A, B, and O blood groups, a discovery that led eventually to the establishment of "blood depots" for British soldiers during the First World War. Advances in transfusion technology permitted more than thirteen million units of blood to be used in transfusions during World War II.

Although the cells I would be receiving would be selected from circulating blood, the procedure was still described as a transplant. This was the terminology used by Dr. Don Thomas long before the existence of stem cells was known, when he took whole bone marrow from a donor and gave it in a recipient. The history of taking a whole organ from one person and giving it to another goes back even further than blood transfusions, to the fifth century B.C., when a Chinese physician supposedly exchanged hearts between two men to balance their disruptive personalities. By the second century B.C., surgeons in India were successfully performing skin grafts to reconstruct noses cut off to punish criminals or lost in swordplay. In most cases autografts (skin moved from one place on a patient's body to another) were successful, while allografts (transplants from a donor to a patient) almost always failed.

Teeth were being transplanted by the seventeenth century. By the middle of the 1800s, corneal transplants were a common procedure. Though there had been many attempts at whole-organ transplants, it was in 1902 that doctors in Vienna successfully transplanted a kidney between animals. The first human kidney transplant took place in 1933, although the kidney never regained function. In 1954, Dr. Joseph Murray, working at Brigham Hospital in Boston, successfully transplanted a kidney from one twin to another. Among the researchers assisting in that operation was Don Thomas, who, two years later, performed the first successful bone marrow transplant. Thomas moved to Seattle to continue his transplant work, and was instrumental in founding the Fred Hutchinson Cancer Research Center.

The Hutch had established itself at the leading edge of transplant technology. The great difficulty that had to be overcome in all types of transplants, whether of an entire organ or of a thimbleful of blood, was immune-system rejection. Stem-cell purification, T-cell (the infantry of the immune system) depletion, and tumor-cell depletion were potentially major weapons for transplanters to add to their arsenal. All of the knowledge that had been gained through recorded history, all of the experiments and all of the people who had risked their lives, had brought medicine to this point. Now it was my turn to take the next step forward.

Once I'd made the decision to have a stem-cell transplant, several other decisions had to be made almost immediately. Was I going to have an autologous transplant, meaning I would donate my own stem cells, which would be returned to my body after chemo and radiation; or an allogeneic transplant, a procedure in which I would receive purified cells from a donor? Was I going to use the traditional bone marrow as the source for my stem cells, or would they be concentrated from my peripheral blood? What chemotherapy regimen was I going to follow? What radiation therapy? Ollie was able to offer advice, but these were my decisions to make.

An allogeneic transplant—in which I would receive donor cells—would eliminate the possibility that my own tumor cells would be put back into my body, but would expose me to potentially fatal graft-versus-host disease. If the Ceprate system and the purging column could successfully strain out the tumor cells, an autologous transplant would be much safer. And if I didn't have confidence in CellPro's technology, how could I expect that from anyone else? So I decided I would have an autologous transplant. My own blood would first be purified by the Ceprate device, then the tumor cells would be picked out of that purified suspension like picking rotten apples out of a basket, and my stem cells would be returned to my body.

The transplant would be only as effective as the chemo and radiation. If those protocols failed to kill the tumor cells that remained in my body after I'd made my blood donation, the entire process would be a waste of time and resources. Normally, transplant patients receive total-body irradiation. Doctors put the patient under a big cobalt gamma radiator and just radiate the hell out of them. The problem is that radioactivity makes no distinction between healthy cells and tumor cells. Billions of cells die. So even if the patient survives, the radiation causes tremendous collateral damage to the internal organs. Many patients suffer lung and kidney problems for the rest of their lives.

But there was an alternative. At the University of Washington, Ollie Press was experimenting with high dose radio-labeled antibodies, a program that used monoclonal antibodies to deliver radioactive bombs directly to tumor cells. This project had started more than a decade earlier. In Ollie's original experiments, these antibodies carried lethal toxins; theoretically they would bind to all B-cells, which included cancer cells, and destroy them. Theoretically.

The results were tantalizing. Patients went into remission for weeks, sometimes months. The antibodies found their targets, but the toxins just weren't potent enough to kill them. So, in 1986, investigators at the Hutch, among them Fred Applebaum, Paul Martin, and Ollie, began attaching radioactive isotopes to these antibodies, which would then deliver these miniature radioactive bombs directly to tumor cells and blow them up. In addition, these isotopes created a crossfire that killed surrounding cells.

In mice, radio-labeled antibodies safely delivered from ten to one hundred times more radiation directly to tumor cells than to healthy cells, although in human trials it only delivered from two to ten times—but that was still much better and safer than external radiation.

Unfortunately, it is an expensive and time-consuming process. Patients have to be isolated in a lead-lined room in which everything they touch—everything—has to be protected. Even the floor has to be covered with special plastic. Patients are literally radioactive for more than a week. During that period they can have absolutely no contact with other people, and everything they touch, everything that goes into that room, has to stay there.

UW Medical Center is one of the few places in the world where this experimental process was done routinely. But the results had been very good. After five years the survival rate of terminally ill patients treated with radioactive antibodies, prior to stem-cell reinfusion, was almost 50 percent.

But it was still considered an experimental process. It cost about $20,000, and my insurance company refused to pay for it. That meant I'd have to pay for it myself. I was certainly willing to do that—if it really contributed to a cure. I cornered Ollie in a hallway one afternoon and asked him bluntly, "Come on, Ollie, tell me, what do you think? Is it worth it? Are you a believer?"

Ollie rarely gives a definitive answer to questions like this. Generally he explains the options and allows the patient to make his own decision. Not this time. "I'm a believer," he said flatly.

That was good enough for me. I was going to receive the most advanced—and experimental—treatment medical science had to offer. I was going to have an autologous transplant using stem cells purified by my own company's technology. After my stem cells had been selected from my peripheral blood donation, any remaining cancer cells would be extracted from that collection using a B-cell depletion process. After donating my blood, I would go through chemotherapy. But in addition to the chemo, I was going to receive high-dose radio-labeled antibodies to destroy the cancer cells inside my body.

This was experimental medicine. No one, absolutely no one, had ever received this combination of treatments. I just happened to be in the right place—although I wasn't there at precisely the right time.

In medicine, survival is so much an accident of time. Just a year before Alexander Fleming discovered antibiotics in 1928, thousands of people died of bacterial infections. A year later, once-deadly diseases had been reduced to temporary illnesses. In my lifetime I had recovered easily from diseases that had killed tens of millions of people: influenza, bacterial infections, high fevers. Had I been diagnosed with lymphoma only a few years earlier, my disease would have been terminal; I would have had no hope for survival. A few years in the future, I'm convinced, purified stem cells will be used as commonly as penicillin; there will probably be a broad range of treatment options for my disease, possibly even a vaccine against it. I was caught on the cusp of discovery. Ollie had been experimenting with high-dose radio-labeled antibodies for ten years. He had used six different antibodies and radioactive isotopes; he had varied dosage. Some of his patients had lived, and some had died. Failure is as important an aspect of the experimental process as success. With each patient, knowledge had been gained. Whatever happened to me—whether I lived or died—scientists would learn something valuable, something that would contribute to a cure.

Of course, I had every intention of living.

The key to the success of the entire procedure was eliminating the cancer cells from my peripheral blood donation. That would have to be done by our B-cell depletion device. But that device was barely beyond the planning stage. If I expected to have my transplant within three months, somehow it would have to be ready. It was an almost impossible task. There was no time to do the usual preparatory work. We weren't going to cut corners, we were going to slice them right off. This was going to be seat-of-the-pants science.

Joe Tarnowski had been the first person at CellPro I'd told. Without his active involvement, there was no chance the B-cell depletion device would be ready when I needed it. After reminding me that at best this device was going to be held together by chewing gum and baling wire, he sighed deeply and asked, "Can I tell Nicole? We're gonna have to get more people involved right away."

"Sure. Yeah. Eventually everybody's got to know."

Joe immediately walked down the long hallway, past the laboratories, past the storage area, past the common rooms, to the office of Nicole Provost to give her a new assignment: Save the boss's life. Please.

Nicole Provost is a no-nonsense, get-it-done person, the perfect scientist to be in charge of a crash project. Joe and Nicole normally met once each week to review Nicole's current projects. She was working her way through the mountain of paperwork on her desk when Tarnowski knocked on her door. "Can I talk to you for a couple of minutes?" he asked.

"It's not a real good time, Joe," she said. "Can it wait a little while?"

"This is pretty important," Tarnowski said.

Nicole invited him into her small office.

"Look," he began, trying to find the right words. "Look, Rick needs a favor. I know you know about his health problems. . . . Well, he's not responding to chemo. They just changed his diagnosis to mantle cell."

Oh, jeez, she thought, *there's no way out of that one.*

"Rick and his doctors have decided he needs to go for a transplant," Joe continued, then added casually, "And he wants to be purged. He wants to use our prototype."

Nicole was incredulous. "There's no way, Joe," she responded, echoing Tarnowski's thoughts. "He can't. It's not ready. We're not ready for anything like that."

"I know, I know, but here's the time line we're looking at. He needs it in about eight weeks." Joe paused, then added softly, "Rick's not going to tell you that you have to do this. If you tell him you can't, he'll accept that. But if there's any way you think you can, he'd really like you to try."

Nicole didn't say that it would require a minor miracle to be ready in eight weeks, although that was exactly what she was thinking. Getting even a makeshift device ready to purge in less than nine months would be a spectacular feat, but eight weeks? "I don't know," she said. "That would be . . . I mean, we'd really be stretching it."

But, just like Joe, just like almost all of the people who worked at CellPro, she immediately switched into the can-do phase. *If* quickly became *how.* "I guess we could put three people on the project. I mean, there's no way it's going to be clinical grade. . . ."

She sighed, "Let me just think about it over the weekend. I'll take it up with the group on Monday. I'm not going to force this on them, this is going to be very stressful. But if everybody agrees, then I guess we'll figure out something."

I was very comfortable placing my life in the hands of Nicole Provost. Although we were not social friends, I'd known her since she began working at CellPro as a polymer chemist in late 1991, several months after I joined the company. Her job then was to figure out how to attach protein molecules to a surface, how to get the avidin to stick to our tiny plastic beads.

Nicole was CellPro's earth mother, the kind of person on whom you could depend in the direst of circumstances. Dedicated to her family and active in her community, she really worked at making the world around her a better place. She was a handsome woman in her mid-thirties who wore no makeup and dressed conservatively. Nicole had been at the center of one of the significant tragedies and triumphs in CellPro's brief history, and no story can better describe her qualities than this one.

She became pregnant in the summer of 1992. We were still a small, close-knit company then, small enough for each of us to know just about everyone else on a first-name basis, and this was a big event for us, our first baby. We had no corporate policies to follow concerning maternity, so we sort of followed Nicole's lead in figuring out how to proceed. She had a perfectly normal pregnancy—Nicole glowed for months. At a company Cinco de Mayo party in May 1993, she began having labor pains. Three days later she gave birth.

The delivery turned into a disaster. The baby was born clinically dead. Doctors managed to revive him, and he struggled to live for five days. And then he died.

Nicole and her husband, Frank, and everyone at CellPro were absolutely devastated. I remember how unusually quiet it was. Normally, when I walked around the building, I'd hear lively conversations punctuated often by bursts of laughter. CellPro sometimes seemed alive with the excitement of success. But not then. For several days a deep depression descended on us. The whole place was in shock. This was a death in our family.

Later, Nicole's doctors reviewed the entire labor and delivery to try to determine what had gone wrong, why this baby had died. When they reached the conclusion, the primary physician asked Nicole, "Are you sure you want to know what went on?"

"Of course I want to know," Nicole replied bluntly. "I'm a scientist."

The baby had been asphyxiated. It appeared that his death had been preventable. Had he been delivered hours earlier by C-section, he might have survived. It was a mistake, a tragic mistake.

Obviously, Nicole and Frank could have brought a malpractice lawsuit against these doctors. If they had, they might have received a large sum of money. But they didn't, deciding that a lawsuit wouldn't help anyone and would not bring back their baby.

Nicole showed us how to deal with this tragedy in a most unusual and courageous way. She became pregnant again—and insisted that the same doctor who had delivered her first baby treat her throughout her pregnancy and then deliver this child. Going back to this doctor, Nicole believed, was the most difficult thing she had ever done in her life. But for her it was a necessary part of the healing process, the only way she knew to deal with her anger and bitterness.

This time, she knew, this time the doctor was damn well going to pay attention to her baby.

A healthy baby boy, Trevor, was born at the end of October 1994. In March 1995, Nicole returned to work. So she knew about life and death. She was a compassionate pragmatist. Most people don't get to choose a person to save their life; I did. And I felt she was a fine choice for that role.

This wasn't a sentimental choice. My life was at stake, and I wasn't thinking about other people's feelings. I wanted the best person possible running this project, and both Joe Tarnowski and I had total confidence in her. If this could be done, Nicole was the person who could do it. In addition to her proven character, Nicole Provost possessed those qualities necessary in a good scientist: intelligence, creativity, self-confidence, tenacity, and good common sense.

Like so many other scientists I've known, she had not grown up intending to become a scientist. Nicole was the daughter of a chemist. She was raised in a Washington State town so small it didn't even have a name, in an area known as Schneider's Prairie. It was a good place for a curious child to grow up, she remembers, the kind of place that subtly molds a personality. As she said fondly, "We always had the space and time to poke around."

Her parents encouraged her natural curiosity. For some reason she has never figured out, in fifth grade she became fascinated with Braille—although she didn't know a single blind person. Her parents bought her a Braille typewriter, and she learned the language of raised dots.

Apparently, Nicole was the kind of kid who loved mixing things together just to see what would happen. Her father would bring home chemicals from his laboratory and allow her to play with them. It wasn't science, it was playing. She'd mix whatever she found in the refrigerator with her mother's cosmetics to see what colors she could create or find out what would bubble or fizz or hiss. When her parents bought her a chemistry set, she didn't like it at all. It came with instructions, the results too predictable; she much preferred mixing her own concoctions.

Nicole loved chemistry. Chemistry was creative; she could throw just about anything she wanted into the pot to see what happened.

That same curiosity made her an experimental cook. Cooking became Nicole's creative release. Cooking and chemistry, she discovered, required very much the same approach, the mixing of different ingredients to make something new. Her mother made sure Nicole and her sister, Ramona, knew how to cook. "I experiment in the kitchen," Nicole admits. "I don't usually follow a recipe too closely. To me the challenge is, how can I alter some things to make them measure up to what I've conceived in my mind? It's a nice exercise to be able to plan out a meal and get results in my lifetime, because in scientific research that is not often the case."

There was a time when the laboratory, like most other areas of American society, was dominated by the man in the white lab coat. All the advertisements, all the annual reports, included photographs of distinguished-looking men standing in front of a slightly out-of-focus lab table on which were visible rows of test tubes and maybe a microscope. It was always men. The presence of these obviously dedicated men was supposed to inspire confidence in consumers and investors. The great influx of women into leadership roles in the laboratories began with the growth of research dollars; suddenly there were more positions to fill than qualified male Ph.D.s to fill them. Nicole grew up in the middle of the women's liberation movement. She was a junior high cheerleader until it occurred to her that it was ridiculous for women to stand on the sidelines cheering for men. Then she set her own course.

After two years at Wellesley, an all-women's college, in 1978 Nicole took a leave of absence to figure out what she wanted to do with her life. While making that decision she began working as a technician in a hospital in Boston, processing blood samples and conducting simple analytical tests.

It was there that Nicole realized how much she enjoyed working in a laboratory. While many labs look surprisingly similar, the reality is that every lab is unique. Each has its own legends and lore, language and personality. A lab very much reflects the personality of its director, the person who runs it. In some labs the director creates competition between members of his or her staff, and makes it a difficult and unpleasant place to work; in others the director is completely supportive of his or her people, and makes it fun to come to work in the morning. Much of the work done in a lab is excruciatingly repetitive; a vast amount of time is spent doing drudge work, counting, loading, cleaning, waiting for reactions, so people do whatever they can to break up the monotony. In the lab in which Nicole worked, for example, she remembers that the technician responsible for measuring the volume of urine in a patient sample would sing loud and happily whenever a new specimen arrived, "Urine the money . . ."

It was while working in this lab that she realized how much she loved the puzzle-solving aspect of laboratory science. Why does this happen? What does that mean? What if we do it this way? It was research that caught her fancy, rather than working in a clinical lab. Doing research gave her the opportunity to explore the limits of her own ingenuity, and she took to it fast.

After graduating from Wellesley, she obtained her master's degree in bioengineering from the University of Washington. At UW she studied the biocompatibility of blood and various polymers, trying to figure out what causes blood to stick to some materials but flow freely through others. Eventually, Nicole decided to specialize in neurobiology, the study of the nervous system, and went to work as a postdoc in a neurobiology fruit fly lab at Brandeis.

Nicole had married Frank Fay, a computer software engineer, while still in grad school. In the 1990 recession, Frank lost his job and Nicole had a big decision to make. "I looked around at all the people I knew in academia, and very few of them were really happy," she remembers.

"Between the crazy hours, the low pay, the precarious grant situation, and the abundance of neurotic co-workers, it just didn't seem like that was going to work for me. While I had really enjoyed working with fruit flies, money had become an issue. Fruit flies weren't going to pay the bills."

The secure jobs and the good salaries were in biotech, specifically in immunology. When Frank got a new job in Seattle, Nicole joined him there and quickly found another postdoc position doing genetic research in an immunology lab. In early 1991, a biotech job-placement counselor called Nicole and told her that a Bothell, Washington, biotech start-up named CellPro was looking for a chemist with a bioengineering background, someone who had experience attaching proteins to polymer surfaces. Nicole applied for the job—and was quickly turned down.

Several months later, after we'd failed to find someone with more experience to fill that position, Nicole was invited to make a presentation at CellPro. Though it's referred to as a presentation, it's actually a sales pitch, and the product you're selling is yourself. Basically, Nicole had to stand up in front of the people with whom she would be working and prove to them that she was perfect for the job.

Nicole was ambivalent about making this presentation. Shortly after moving to Seattle, she had joined several groups to learn more about the local scientific/biotech community. Among those groups was the Association for Women in Science. Nicole and several other women with whom she'd gone to school at UW had formed a little splinter group, DWIS: Disgusted, Determined, Disenchanted, Decisive Women in Science. Essentially it was a support group for female scientists considering a career outside academia. When Nicole was asked to present at CellPro, she hesitated. It had been almost a decade since she'd worked in polymer chemistry at UW. She thought that pretending she knew more about polymers than she really did was intellectually dishonest, maybe even unethical. She hadn't done polymer chemistry in so long that she doubted she was even qualified for this job.

But then she began wondering, *What would a man do in this situation?* And she decided a man would say, "I'm qualified," and go after the job. After reviewing her almost decade-old master's thesis, she spent the next four days refreshing her knowledge of polymer chemistry.

All of which brought her to CellPro, and the challenge of a lifetime. She was back in the business of blood.

I don't remember the first time I met her. I'm sure we were polite with each other; we probably shook hands and nodded when we passed in the hallways. CellPro had just gone public, and we were beginning to expand. Nicole was one of our first new hires. When possible, I like to spend time nosing around in the labs, just to keep up to date about what's going on. But we were so busy building a company that I rarely had time to do that. My only real contact with Nicole came at our monthly staff update meetings. It was obvious she was a fine chemist. By 1996, Nicole Provost was a senior scientist at CellPro, meaning she had overall responsibility for several projects. Among them was the development of the tumor-cell depletion system for breast cancer patients.

After meeting with Tarnowski, Nicole asked three research associates, Sharon Adams, Kirsten Stray, and Stan Corpuz, to join her in a small conference room. Kirsten and Stan had been working on the breast cancer project, while Sharon had been working on our recently concluded attempt to harvest fetal stem cells for genetic testing in place of amniocentesis.

Nicole met with her people at least once every week, and usually more often, to review progress and plan strategy. But this meeting was very different. After conducting normal business, Nicole pushed aside her paperwork and said, "Okay, we've got a decision to make. Rick needs a favor. He wants us to speed up the lymphoma project. They've changed his diagnosis to mantle cell, and he's not responding to chemo."

Nicole paused to let that news sink in. Nobody said a word. Four months earlier, Sharon Adams and Kirsten Stray had attended a symposium on mantle-cell lymphoma at the annual meeting of the American Society of Hematology. It was a disease that affected men almost exclusively, and the prognosis was very bad.

"It's even worse than that," Nicole finally continued. "He's going to have a transplant, and he wants to be purged, and he'd like to be the first patient to use our system." Finally she added, "And he wants to do it in eight weeks."

Sharon and Kirsten moaned. But Stan Corpuz brightened. "All right!" he said enthusiastically. "Let's go for it."

"I haven't told him we would do it," explained Nicole. "Look, I don't want to go forward with this unless we all agree. It's your choice. I promise you, if you don't want to do it, it isn't going to affect your jobs.

But if you do, you need to know it's gonna be a really hard slog. It'll mean that the three of you have to work together as a team. It's probably going to mean working a lot of nights. You're gonna have to take responsibility for it. I'll help organize it, but I don't work in the cell lab. You guys have to call the shots."

This was Sharon Adams's first day back from a wonderful vacation in Mexico. She was tanned and relaxed. Suddenly she could feel the pressure descending over her like a dark blanket. "It's your choice," Nicole had said, but Sharon knew there was no choice. She knew it would be difficult, but she was confident it could be done. "Well," she said softly, "the good news is that if we don't get what we need, we really can go right to the top."

They were in that room for slightly more than a hour, their initial surprise growing slowly into excitement. This was the kind of in-your-face challenge that scientists rarely confronted: develop a potentially lifesaving technology in eight weeks. Nicole didn't tell them that this project was to be kept secret, but they agreed to keep it low-profile. They had to have something to call it among themselves, though. The project had to have a name. While it was a derivative of the breast cancer purging project that was limping along, it was much more than an extension of work already done. They had to choose an antibody and maybe even a different cell-selection system. But all the work had a single objective: saving my life by eliminating every tumor cell from my stem cell harvest. There was really only one perfect name for this work, and so it became known as the Rick Project.

Nicole Provost had taken charge of the breast cancer depletion project after returning from maternity leave. At that time it was a project going nowhere slowly. We felt certain we could produce an effective tumor cell-purging column; what we didn't know was how to get it approved by the FDA. We had to demonstrate that it had some clinical value, which was almost impossible. We feared that if we tried to get it approved, the FDA would insist we prove that eliminating tumor cells improved survival rates. Even if we could do that trial, it would take at least five years to compile sufficient data.

We had begun laying the groundwork for a tumor-purging system in early 1993 when a project team had been organized to begin the preliminary work. Among the initial decisions made by this group was that they would focus on breast cancer cell depletion, and that rather than using our avidin and biotin system in both the initial stem cell purification phase and the secondary tumor depletion phase, it would be much better to use a magnetic bead selection system in the depletion phase.

Magnetic beads work in a way similar to avidin and biotin. Rather than attaching biotin to the antibody, a sliver of iron less than one-twentieth the width of a human hair is chemically bonded to it. And instead of a column filled with avidin-coated beads, there is a large magnet. When the mix flows through the column, the iron is attracted to the magnet. It's a very strong connection. The use of magnetic beads is an excellent technology; it'll produce purities of more than 90 percent in small-scale experiments, and has been used to target and select cells since the early 1960s.

The most serious problem with magnetic beads is that they bind firmly to the stem cells, making it very difficult to get them off. That's very desirable if you intend to discard the cells stuck to the magnet, as you would in a purging system. You throw out the column with the selected cells firmly attached. But it's not great if you need to use those cells again, as you do when collecting stem cells for a transplant. The beauty of the Ceprate system was the ease with which stem cells were released from the column after being selected. The flexible polymer beads we used just sort of released the cells when mildly agitated.

Baxter used magnetic beads in its Isolex system and had trouble liberating the magnetic beads from the stem cells. They had tried several different approaches before finally using synthesized peptides as a releasing agent. The peptides successfully broke the bond between the antibody-bound bead and the cell. Unfortunately, a new peptide had to be synthesized for each different antibody, making this process extremely expensive and severely complicating Baxter's ability to adapt the Isolex to new applications.

In our system the tumor-cell depletion would be a second step. Using avidin and biotin in both the selection and purging steps would create a real chemical mess, requiring the addition of blocking agents at different points.

So going to a magnetic-bead system to fish out tumor cells made sense. But when the group made that suggestion to our vice-president of engineering, another ex-Baxter employee, he rejected it immediately. There were several good reasons for that. For us, this was an entirely new "catch" system. We were committed to avidin-biotin, and we probably knew as much about it as any company in the world. Changing to mag-netic beads would require finding new suppliers and investing consider-able resources in a technology we didn't really know very well—but certainly a primary reason for rejecting it was that Baxter was using it in its Isolex system. And we just didn't want it to look as if we were copying Baxter. The group was told to find another way to deplete tumor cells. So the project pretty much languished on the shelf.

But transplanters continued to tell us that they wanted a tumor-cell depletion system. They saw their patients relapsing and dying, and be-lieved the fact that they were reintroducing tumor cells into their bodies may have been contributing to that. When it became clear at the end of 1995 that our attempts to select fetal cells from maternal blood and use them as an alternative to amniocentesis was not progressing, we ended that project and once again began looking seriously at a tumor-cell purg-ing system. There was obviously a market for that product. So we began some rudimentary experiments, just sort of feeling our way around the area, trying to gain some understanding of what we wanted to do and how we wanted to do it.

And then I received the e-mail from Ollie Press. Having been diag-nosed with mantle cell lymphoma certainly changed my motivation. As the CEO, I wanted to do what was best for my company. But I wanted to live; more than anything, I wanted to live. I was also convinced that producing a tumor-cell depletion device was a legitimate project for this company.

Every scientific project begins with the identification of a need. But rarely, if ever, is the need so personal. We had been aware for several years that there existed a potentially large market for a tumor-cell depletion sys-tem. We just never expected to need it so soon, nor did we anticipate that the stakes would be quite so high.

Although Kirsten and Stan had been working on the breast-cancer depletion project for several years, they really had to start this project from scratch. While they'd made significant progress, it was not readily adaptable to the Rick Project. Breast cancer and lymphoma are very differ-ent diseases. In lymphoma, the patient's circulatory system is overrun by a huge number of malignant B-cells, while breast-cancer patients typically have only a small number of tumor cells circulating in their blood. The prototype column they had been using to deplete breast-cancer tumor cells was just too small to process the millions of malignant cells present in lymphoma. A new, larger column had to be developed. Two new anti-bodies had to be selected and tested. Flow rate—the speed at which the blood product passes over the column—had to be determined. A decision had to be made about whether to do the tumor-cell depletion phase before or after stem-cell collection. And, just as the team had feared, the fact that avidin and biotin would be used for both stem-cell collection and tumor-cell depletion created tremendous problems.

The team wasn't even sure the system would work. In tests, they had successfully detected and depleted breast cancer tumor cells in a labora-tory setting, but the system had also removed too many stem cells from the blood product, making it impossible to proceed with the transplant. They had also experienced problems in measuring accurately the number of residual tumor cells left in the mix. All they knew for sure was that they could take out most tumor cells from a stem-cell harvest; what they did not know was precisely how many tumor cells remained.

All they had was a prototype of a system that had successfully de-pleted breast-cancer tumor cells. In some experiments they had success-fully removed five logs, 99.999 percent of all detectable malignant cells—enough to satisfy physicians in real-world settings. So they knew it worked—at least on a laboratory scale. It needed a lot of jiggling—a bit of this, some of that, a dash of something else—but they had proved that the theory could be transformed into tumor cell depletion. They had proof of principle.

The x factor was the clock. In the broadest sense, the experiment was analogous to the race to the moon: the objective was clearly defined, it had to be accomplished in a defined amount of time, and it had the full support of management. But there any comparison ended. Doing this experiment following good scientific procedures was impossible. That would require months and years, not the days and weeks that we had. So, rather than progressing step by step, carefully examining each small exper-iment to understand the results before taking the next step, they took the most aggressive approach possible. They were going to follow their hunches, pursue what worked, and abandon what failed. The object of this project was to get a depletion system that worked; understanding why it worked could come later.

While the team was getting organized, I began preparing for my transplant. To save my life, Ollie Press and his team were going to come as close as possible to killing me. Unlike whole-organ transplants, a stem-cell transplant actually takes several months. It would begin with the insertion into my body of a Hickman catheter, a flexible tube almost eighteen inches long that would remain there for the entire duration of the transplant procedure. The catheter would stick out of my chest just above the monograms on my shirts, like a hose attached to a fire truck. Without the Hickman catheter, the thousand or more needle sticks I would have received instead would have destroyed my veins. The tube was a tunnel directly into my blood system, a simple and reliable way to get everything that I needed, all the drugs and chemicals and eventually my concentrated stem cells, into my body and take out blood samples.

At home, Patty and I didn't speak about the transplant any more than absolutely necessary. But when we did, it was in the most normal terms possible. Make a dental appointment, bring the car in for a tune-up, go in for chemotherapy. As long as Patty and I had been together, it was still difficult for me to guess what was going on in her mind. Behind our house was a sprawling backyard, around which Patty had built a border of trees, shrubs and flowers—with no help from her family, as she had often pointed out. Patty is a person who needs bright colors in her life. She knew by name several of the squirrels who lived in our yard, and some mornings we would watch them stealing food from Patty's birdhouse. I noticed these days that she spent more time than usual in the garden, pulling out weeds and leaving only her blooming flowers.

That seemed like a perfect metaphor for what Ollie Press and his team and Nicole Provost and her team were about to do inside my body.

I tried to treat this procedure as my job. During this period I had not missed a single day of work, and I was proud of that. I intended to keep working until I was no longer able to do so. Plans had already been made for Larry Culver to run the company in my absence.

I rarely thought about dying, as if even considering the possibility somehow opened the door to my mortality. Instead I focused firmly on the future. But the reality of the situation was unavoidable. I had a disease that is most often terminal. The first effort to save my life, chemotherapy, had failed. I had one chance left, a transplant.

I wasn't afraid of death. I intended to fight for my life with every resource I had available, until I just ran out of time and options. But the one thing I did fear was a decline in the quality of my life until I was no longer capable of functioning as a human being. I didn't want to end my life in pain, as a burden to Patty and my sons. That I couldn't bear. As we drove home from the hospital after a treatment one afternoon, I told her, "You know, if I start to get like that, I think I'd rather go take a long sail on the boat and not come back."

She glanced at me as if I were crazy. "What are you talking about? That's just silly."

Even now I believe I meant it. As much as I worried about myself, I was far more concerned about the toll it would take on my family to watch me die slowly. "Well, you know," I said, about as offhandedly as I could manage, "if we reach that point and I'm coming close, I'd rather cut it short." If all hope was gone, I wanted to get into my boat and sail into the ocean. Eventually they would find an abandoned boat at sea. That would be so much easier for everyone. And for me that would be a peaceful way to end my life.

But that was still a great distance in the future. I had a battle to fight.

The first step in the transplant process was to extract as many stem cells as possible from my blood. If we couldn't collect enough stem cells to proceed with the transplant at the very beginning, that would be the end of the procedure. The minimum amount of stem cells needed was based on the weight of the patient. Stem cells compose such a small percentage of blood cells that it is extremely difficult to identify and collect them.

At one time, stem cells could be extracted only from bone marrow. To collect enough marrow, donors had to be poked in their hip bones with a long needle hundreds of times, an expensive and painful process. But that changed with the discovery that the immune system could be stimulated to produce stem cells by drugs known as growth factors, or by growth factors combined with chemotherapy, and to send them out of the marrow into the peripheral blood system in great numbers.

After the body has mobilized, sending billions of stem cells into the circulating blood system, the cells are removed by a process called apheresis, a Greek-derived word meaning "take away." Apheresis is very different from an ordinary blood transfusion or donation. The donor is hooked up to a large instrument that looks very much like a dialysis machine. Blood is taken out of a vein, just as in a normal donation, but, rather than being collected in a big bag, it flows into the apheresis instrument, where its components are separated by centrifugation according to their density. The plasma and red blood cells are returned to the donor, but a light pink solution of white blood cells is collected in a pouch. Inside that pouch are billions and billions of white blood cells, among them the stem cells that might save my life.

That solution was the raw material for Nicole's experiment.

Several weeks after this procedure, I would be given high-dose chemotherapy, which would destroy all the blood cells remaining in my body—including the tumor cells. And then I would be given back my stem cells. If Nicole's team was successful, I would receive pure stem cells. It would take slightly less than two weeks for these stem cells to grow into a new blood and immune system. If the experiment was not successful, in addition to stem cells I would also receive tumor cells. Gene-marking experiments had shown that tumor cells returned to a patient in a graft could contribute to a relapse—meaning these cancer cells would also engraft and grow, and eventually my cancer would return.

I tried very hard not to interfere with Nicole and her team. Instead I concentrated on CellPro business. And business was pretty good. We were selling the product very successfully in Europe. The clinical trials in which we were involved continued to show tremendous potential. Our stock remained steady, and brokers and analysts were advising their clients that we were a good buy. With all this going on, I was not even aware that in late April 1996, the United States Supreme Court made a decision in a dry-cleaning dispute that would impact directly on the existence of CellPro.

A man named Herbert Markman had invented and patented a data-processing system that used a bar code to track individual articles of clothing through the entire dry-cleaning process, from the moment they were handed over to a clerk until they were ready to be picked up, and prepared a bill. According to the patent, this system was able to "maintain an inventory total" and "detect and localize spurious additions to inventory." A product created by a competing company, Westview Instruments, also used a bar code to calculate dry-cleaning charges, but it was not capable of tracking individual articles of clothing.

Markman sued Westview, claiming the word "inventory" included "articles of clothing," or "dollars" or "cash" or "invoices." Westview's defense was that since its system was not capable of tracking individual pieces of clothing, it did not infringe the patent claim.

After hearing the evidence, a jury decided that Markman was absolutely right: Westview's system violated his patent. But a district court judge disagreed. He overturned the jury's verdict and issued a verdict from the bench for Westview, pointing out that Westview's system was unable to track the dry-cleaning inventory as that term was defined in the claim—meaning it couldn't track individual articles of clothing throughout the entire process.

Markman appealed, claiming that the district court judge did not have the legal right to substitute his interpretation of the patent for the jury's decision. The court of appeals upheld the district court judge, and the case eventually went to the Supreme Court.

This was the kind of case that fascinated patent lawyers, but almost no one else even knew or cared about it. The real issue was not the intricacies of a dry-cleaning tracking system, but rather the right of a judge to reconstrue a patent claim. Basically, it took claim construction out of the hands of juries and delivered them to judges. It certainly did not seem likely that a case revolving around the meaning of the word *inventory* would eventually affect the lives of so many people.

The American Patent Act does not specify whether the court or a jury has the right to determine exactly what a patent claim means, so the Supreme Court looked back in history to eighteenth-century British courts to answer that question. Basing its decision on English patent law in effect before the United States Congress passed the first American patent legislation in 1790, the Supreme Court ruled that the court, meaning the judge, had the right to interpret patent claims. Part of their argument was that judges were smarter than juries, and therefore "better suited to find the acquired meaning of patent terms." Only after the judge had determined the meaning of a claim did the jury have the right to decide if it was valid or if it had been infringed.

Among those people who may have read this decision with great interest was District Court Judge Roderick McKelvie, who apparently remained quite upset and angry that a jury had found unanimously on 103 charges that we had not infringed Baxter's patents. The *Markman* decision basically put a sledgehammer in his hands, and he set about using it to pulverize our case.

7 THE RICK PROJECT BEGAN WITH THE ARRIVAL OF a bag of fresh blood. People generally associate blood with pain, injury, or disease; they see it only when there is an accident or at a doctor's office or in a hospital. But at CellPro, blood was the currency of our survival.

Blood was very expensive, but the company couldn't live without it. Every experiment we conducted, every test, required a blood product. What made our needs difficult to fulfill was the fact that we couldn't use whole blood, the type of blood donated at blood banks. We needed only that component of blood that contained stem cells, and obtaining it was a lengthy process. We had set up a very sophisticated donor program. After potential donors were tested for an array of blood conditions, they were given a growth factor to mobilize their immune systems. It generally took donors about a week to mobilize. In the hospital, donors were hooked up to an apheresis instrument for about three hours, during which time twelve liters of their blood were processed. They got their plasma and red blood cells back, and we kept the white blood cells. These donors had to return to the hospital for two or three consecutive days to complete the entire procedure.

We paid the hospital $1,500 per unit of blood product, of which the donor would receive about $450 for two collections. The cost of blood ranged between $5,000 and $7,000 per week.

When we couldn't get enough mobilized peripheral blood, most research and development stopped. We were no different from a fleet of cabs during the oil shortage. There were times when everybody in our labs was just standing around waiting for blood to arrive. But from the day the Rick Project began, Nicole and her team got first call on all the blood that came through the door.

Unfortunately, there was no way to increase the flow of blood. We couldn't just turn on a pipeline. Even if we had found more donors, we were limited by the number of available apheresis instruments in Seattle. Most of those machines were being used to support cancer patients whose lives depended on them.

What we had in abundance were brains and sweat. Nicole had assembled an excellent team to attack this project; Stan Corpuz and Kirsten Stray, both of them among CellPro's first employees, and Sharon Adams. Coincidentally, each of them was going through a difficult period in his or her own life when they began working on the Rick Project.

Years ago, when I read about the legendary scientists, from Newton to Watson and Crick, I never thought of scientists as people wrestling with the problems of everyday life. It never occurred to me that Madame Curie had bills to pay, or that Alexander Fleming might have suffered at love.

I've spent a great deal of time in laboratories. I've worked with many, many scientists and medical technologists. And perhaps the most surprising thing to me is how utterly normal they are. The work they do will certainly change your life—it may even *save* your life—but the people who make these discoveries, who do this work, fight the same daily battles as everyone else. They worry about job security and paying the bills; they struggle to raise their kids and try to navigate their own way through relationships; they can be just as nice or nasty as anyone else, just as generous or greedy; they might drink too much or drive too fast. The entire spectrum of human frailties plays itself out on a daily basis in a laboratory. The same Dilbert cartoons that hang on the walls at law firms and shipping companies appear in every lab.

I tried to maintain a business-as-usual attitude during the Rick Project. Although at times I felt like wandering casually down to the lab and asking, "Oh, by the way, how's that B-cell depletion project going?" I managed to restrain myself. Seeing me standing there in my ever-present hat would do nothing but add more pressure. I had confidence that the people down the hallway knew exactly what they were doing.

The people down the hallway . . . For me, this project was the single most important scientific research being done anywhere in the world. For them it was a job. An incredibly difficult job. I had spent some time traveling with Stan, but I really didn't know any of them very well. And until much later I certainly did not know what was going on in each of their lives.

On those few occasions when I think back to this incredible project, I'm astonished by the confluence of events—what many people call fate—that had to occur to put the right people in precisely the right place at the specific time I needed them. The only common bond we shared was our great respect for science and medicine.

I've never found a universal reason why people are attracted to science, but I've noticed it seems to start at an early age, often with a curiosity about the way the world works. If a study were done, I suspect it might reveal that people drawn to science are unusually inquisitive and disciplined, somewhat creative, and have good logical minds and basic mechanical skills.

Stan Corpuz followed his childhood curiosity into a career in science. Stan had been one of CellPro's first five employees, hired by Ron Berenson when the company was building the prototype cell separation system in the converted morgue. Stan was a Hawaiian of Filipino-Spanish ancestry. Only five feet three inches tall, he made up in determination what he lacked in size. For example, he had celebrated his fortieth birthday by successfully completing his first triathlon.

After graduating from college with a degree in biology, he'd worked at the Puget Sound Blood Center, coincidentally only a short distance from the lab in which Patty worked. Eventually he became a lab technician at the Hutch, working with scientists trying to determine the role played by neutrophils—part of the white blood cell family—in the circulating blood system. For several years he worked primarily with rabbits—and he realized it was time to find another job when in his dreams he saw a very large rabbit coming after him with a syringe.

Stan had met Ron Berenson through his roommate, who was then working in Berenson's lab at the Hutch. When Berenson began putting together the lab staff for his new company, he needed a technician with a range of experience and skills. The fact that Stan had been working with the avidin-biotin technology at the Hutch made him the perfect choice.

Stan and I had taken several business trips together, and I remember him telling me about the early days of CellPro. "Back then, at the beginning, none of us had any life except the lab. Sometimes we worked all day and all night. If blood was coming in at night or on a Saturday, we had to be there to work with it.

"People kept telling me we were going to fail, but we believed in what we were doing. It was impossible to spend time with Ron Berenson and not believe it. Maybe it was just the competitive nature in each of us, but we really wanted to make it work, we wanted to get it out of the lab and into the clinic for testing."

Stan had become our office optimist. No matter how difficult the situation, he was always upbeat, always certain it would work out. But at the beginning of the Rick Project he seemed unusually subdued, even sullen. Normally gregarious, he walked around the lab like a lit fuse, ready to explode at the slightest disagreement. This was totally out of character for him. For a long time he refused to talk about his problems, but finally he admitted to Sharon that he was having a very difficult time at home. Just before the Rick Project began, the woman to whom he had been engaged for two years, a woman he loved deeply, had broken their engagement and moved out. He was going home each night to an empty house that still resonated with the sound of her voice. Too many nights he found himself falling asleep on the couch in front of the television set. He was struggling to deal with a broken heart.

The days were not much better. The excitement he'd once found in the lab had disappeared. The challenges were gone, and the work had become endlessly repetitive. It didn't seem to have a whole lot of meaning. It was a job and nothing more. Stan had lost contact with the purpose of his work. In all the important parts of his life, he was desperate for change. For Stan, the Rick Project offered a chance to emerge from his depression by doing something important for someone else.

In the spring of 1995, Kirsten Stray's husband had been accepted into graduate school in Utah and had moved there, leaving her alone in Seattle with a house, the dogs, and the responsibilities. As often as possible she would fly to meet him for the weekend, or he would return to Seattle, but basically they were separated.

Kirsten often played the role of quiet peacemaker between Stan and Sharon Adams, both of whom admit to being stubborn when they feel they are right. Like Kirsten, Sharon was often alone. Her husband was the operations manager for a construction company, a job that forced him to spend as many as four days each week on the road, leaving her with their twelve-year-old son, Jason. But even considering that parental balancing act, the Rick Project came along at just the right time for Sharon.

Sharon had inherited a love for the laboratory from her father, a high-school chemistry and algebra teacher. As a child she was always leaving experiments around the house, from petri dishes on the windowsills to the partially dissected turtle her mother found unexpectedly in the refrigerator. She grew up believing that someday she might save the planet or cure cancer, and although she eventually lost that youthful idealism, she never lost her enthusiasm for the laboratory.

After spending more than nine years doing immunology research at Stanford, she found a job at CellPro when her husband was transferred to Seattle. As she discovered, there's a big difference between working in academia and in the corporate biotech world. For a long time it had been generally accepted that people who took jobs in industry were selling out, that the only pure scientific research was being done on university campuses. And for a long time that assumption wasn't totally incorrect.

In the world of big-time biotech, the pressure to produce as quickly as possible is tremendous. There isn't time or opportunity for a scientist to follow his or her curiosity. The company is burning money every time it turns on the lights, and if it can't get a product to market, jobs will be lost, careers ended. Without success in the labs, the lights do get turned off.

While the politics in a university lab can be brutal, and students can be exploited by ambitious primary investigators, there is considerably less pressure for technicians. Money and jobs aren't so directly at risk. It's much more an incubator for creativity. Sharon had spent years in a very good lab at Stanford, working for supportive primary investigators on interesting projects. It was an idyllic situation. After all that time, the people with whom she worked had become her extended family.

The transition to industry had been difficult for her. She'd spent much of her time on the fetal-cell project searching for a usable antibody. It was tedious work, and eventually she got very bored doing the same thing day after day. And then the decision had been made to pull the plug. Progress wasn't being made fast enough to justify the investment.

Sharon was a perfect fit for the Rick Project. A B-cell antibody was the foundation on which the whole depletion system would be built— lymphoma tumor cells are typically B-cells, although not all B-cells are tumor cells—and she had spent several years working with B-cell antibodies. Stan and Kirsten could design a workable column, but unless we found an antibody that would seek out and bind to my tumor cells, the system wouldn't work. It would be like using a rocket to launch a colorful beachball rather than a communications satellite. For Sharon, this was perhaps a final chance to find the satisfaction in the corporate world that she'd left in academia.

The team began by building a model system, a scale model of the final device, on which they could conduct experiments and measure results. This was the biological equivalent of a wind tunnel. In addition to the device prototype, they also needed a witches' brew, a model of my diseased blood, to run through it. It didn't have to be my blood, just blood contaminated with tumor cells—which they made. They chemically stained some cultured tumor cells, then added these now glowing cells to a bag of healthy blood. After running this blood through the model system, they would run a sample taken from the collection bag through a FACS instrument, a device sensitive enough to find one fluorescent tumor cell in every 5 million cells. If any tumor cells remained after the blood had been run through the prototype, the FACS machine would detect them.

Selecting the antibody to be used was actually the first step. B-cell antibodies could be grown in our own lab or purchased off the shelf from suppliers. It's not quite the same thing as going to the deli counter and ordering a pound of B-cell antibodies, but they are commercially available. We had previously licensed the rights to two antibodies, meaning we could use them in our products, and the team began running experiments with them.

Given enough time, the team probably could have discovered an antibody specific to my tumor cells, one that would bind to those cells and no others. But we didn't have enough time. Because my tumor cells were simply B-cells gone berserk, a universal B-cell antibody that would recognize and bind to all B-cells, healthy B-cells as well as tumor cells, would have to be used. That was acceptable. I wouldn't need the healthy B-cells that would be eliminated in the process; in fact they were considered a contaminant. The only cells that really had to be put back into my body were stem cells.

The two B-cell antibodies to which we already owned a license turned out to be extremely difficult to grow in our lab. I think the actual scientific phrase Nicole used to describe them was "real stinkers." So we had to find an outside supplier for our antibody. The ticking of the clock was getting louder.

The team just didn't have time to test all the available antibodies to identify the one or two that worked most effectively. We had to settle for whatever we could get right away. We finally got an acceptable B-cell anti-body from a French supplier. Although we could grow it in our lab, the problem with it was that it had not been clinically validated. It was dangerously "dirty."

In this case, "dirty" means possibly contaminated with bacteria, viruses, or endotoxins. Endotoxins are bacterial agents that can produce very high fevers. Dirty antibodies are used almost exclusively for research or diagnostic applications. In those situations a little contamination has no effect. But for a human being, a dirty antibody can be deadly.

Joe Tarnowski explained the situation to me. "The stuff's not clean," he said. "We're going to filter it, and we're pretty sure it'll be sterile, but I want you to be aware that we're doing it in the lab, not in a clean room. There's no way I can guarantee you that it isn't contaminated."

"So? What's the worst that can happen?" I asked, not quite flippantly.

"You might die." These cells were going to be put back into my body after radiation and chemotherapy had destroyed my immune system. My body would not be capable of fighting an infection. Even something as normally harmless as a high fever could cause potentially fatal toxic shock. "They'll pump you full of antibiotics," Joe continued, "but you have to know it's a risk."

I had stage-4 mantle-cell lymphoma. The most optimistic estimates gave me thirty months to live. I had no options left. "Okay," I said, "let's use it."

There was another problem to consider: we did not have FDA approval to use these antibodies in this system. Like all of our clinical antibodies, these had been grown in mice. Hybridomas, the internal factories that mass-produce a desired antibody, are made by the fusion of an antibody-producing B-cell with an immortal cancer cell—resulting in an immortal antibody-producing cell. That cell is then injected into the spleen of a mouse, which turns that mouse into a living antibody factory.

But because there is always the possibility of transferring a disease from a mouse to a human being, the FDA tightly controls the production and use of antibodies. The only way we could use an antibody grown in this way that had not been previously validated, meaning it had been tested and found safe, then licensed for use in human beings, was to obtain what is known as a "compassionate use" exemption from the FDA.

That's essentially a waiver given by the FDA to a doctor allowing an experimental process to be used, based on the theory that the patient is terminally ill and cannot be harmed by this procedure. In these cases the possible benefits far outweighs the risks.

Our Director for Regulatory Affairs, Dr. Monica Krieger, did most of the necessary paperwork, requesting permission from the FDA to use this antibody in an unnamed patient. Dr. Scott Rowley, director of the processing laboratory at the Hutch, where the purification and depletion would have to be done, actually filed the papers. It was all done according to regulation. All the *t*'s were properly crossed and the *i*'s firmly dotted. The FDA had thirty days to respond.

While this is done primarily to protect the public safety, it's also done to protect the individual patient. Terminally ill people are desperate for any hope. When they run out of medical options, some of them begin looking for miracle cures. And there are still snake-oil salesmen selling hope at a great cost. Actually, these people don't sell snake oil anymore; it's usually a "cure" like shark cells, based on the claim that there is something in the immune system of sharks that prevents them from getting cancer. The mountain of FDA regulations prevents seriously ill people from being exploited at the most vulnerable point in their lives. So, if these patients want to spend money on miracle cures, they have to do it outside the United States.

I was certain that the FDA would issue a compassionate-use exemption, but just to hurry along the process, I decided to make my case personally. Speaking with some FDA staff members on a conference call, I said, "I don't know if you're aware that CellPro has filed for a compassionate-use exemption for our B-cell depletion column, but I have to tell you—that's for me. I'm the patient. . . ."

The FDA was aware that I had lymphoma; they did not know that I was going to have a transplant, and they certainly didn't know that we had started working on a B-cell depletion system. An official procedure had just become very personal. This might well have been the first time that a patient directly requested a compassionate-use exemption.

". . . So, you know, I'd really appreciate it if you guys could look at this pretty quickly. We'd like to move ahead with this as soon as possible."

"We'll certainly do that," was the response. The application was approved quietly. Would I have used the system if the FDA had turned me down? Absolutely. I wouldn't have hesitated. Personally, I believed completely that this system would help save my life, and I was willing to take my chances. This wasn't shark cells, it was good science. But, professionally, I wouldn't have had the opportunity. If we had done this procedure without permission, the company would have been fined substantially, and people who participated might have lost their licenses to practice medicine. It's serious business.

While mice are still one of the most accepted and commonly used source of antibodies, scientists are always searching for better and safer methods of producing antibodies. For example, recent experiments have proved that it's possible to manipulate the genetic structure of corn or tobacco to literally grow antibodies. Imagine using fields of tobacco plants to grow antibodies to cancer cells.

After running the initial experiments, the team decided to use two different antibodies, known as anti-CD-19 and anti-CD-20. The reason for this is that antigens—the specific proteins on the surface of the cell that attract and hold antibodies—appear on cells at different times in their life cycle and then often disappear, sort of like baby fat or hair. Using a cocktail of two antibodies makes it possible to capture B-cells at different stages of their maturation. Clean or dirty, we had our antibodies.

For the first few days the Rick Project was like a new engine turning slowly, as the rough edges were worn off the gears until they fit together smoothly, and the engine grew powerful. Nicole met daily with her three research associates to review progress and plan strategy. A tremendous amount of preparatory work had to be done before they could really get down to work. A rudimentary column had to be assembled, antibodies had to be tested, the materials that would be needed had to be organized. There was no way of predicting how many different small experiments would have to be conducted before the system was ready for prime-time use. Hundreds, at least. This was being done on the fly—"Fun with chemistry," as Nicole referred to it.

Normally, the experimental process is like creating a path. You go back over the territory again and again until you, or anyone else, can follow it in the dark. What the Rick Project team was doing was more like building a staircase. They proceeded by trial and error. An experiment either worked or it didn't. If it worked, they would move to the next step. There was no time to discover why it worked. If it failed, they would take a guess and make another attempt.

For the first two weeks the team tried to apply the knowledge and experience gained in the breast-cancer depletion project to this one. Theoretically, it should have worked. But it didn't. The system just didn't fish out the cancer cells from the flowing river of blood. There wasn't any logical reason for that. The same system that had worked reasonably well to purge breast-cancer cells just didn't work on lymphoma cells. In one configuration the system successfully depleted tumor cells, but also plucked out too many stem cells. It didn't leave enough stem cells to proceed with the transplant. In another configuration it didn't take out enough tumor cells. The only progress made in the first two weeks was the realization that they could pretty much throw out everything they'd learned in the breast-cancer program.

Two weeks were gone, six weeks remained, and they had moved farther back behind the starting line.

They were terribly frustrated. But in fact, important progress had been made. They were gradually becoming a team. Through trial and error they were figuring out how to proceed. At first they didn't even speak the same scientific language. They would disagree about inconsequential things: Do you start counting at this point or that point? How do you present data? What constitutes enough success to move forward? How much is enough?

They began meeting every day in the small conference room, sometimes twice a day, to go over in minute detail everything they were doing. Nicole would join these discussions three or four times a week. To become a team, a diverse group needs to find a common ground, something on which they come together, something that becomes the subject of conversation or a shared joke. A team develops its own totems. For the Rick Project team, the first small piece of common ground was Chee-tos.

Bright orange, crunchy Chee-tos. At one of the first meetings, Nicole brought lunch for everyone, which included a big bag of Chee-tos, and nonchalantly tossed it in the middle of the table. Everyone had their data on sheets spread out in front of them. Sharon had never had a Chee-to. She tried one . . . then a second . . . and was hooked. Stan took one, then a handful. The four of them started eating these Chee-tos. "Okay," Nicole said at one point, "we've got to get rid of these Chee-tos right now."

Everyone agreed, then kept right on eating.

That wasn't completely surprising. In many labs the primary diet is junk food. A balanced diet consists of two of everything in the vending machine. Scientists rarely have time to go out for lunch, so they depend on these machines for food.

Chee-tos became the comfort food for the group, a tonic for the relief of stress. Someone would bring a big bag into each meeting, and as they discussed my fate they would happily crunch away. "The, um . . . B-cells . . . (crunch) . . . aren't properly . . . (crunch) . . ." Everybody would leave the meetings with hands stained orange. After a while they joked that they could measure the level of the pressure by the number of bags of Chee-tos left in the vending machine.

It took the team about a week to figure out how best to work together. These were three bright, independent people, used to generating their own ideas, and each of them had an idea how the three of them should work together. Scientists tend to be people who have to believe strongly enough in their own ideas to pursue them when others express doubt or even disdain. The first thing the group had to learn how to do was compromise. And naturally they each had their own ideas about the proper way to compromise, too.

Stan, Kirsten, and Sharon agreed that each of them would have an equal say in every decision. When they couldn't reach a compromise, Nicole would intercede and make the decision. Nicole was the perfect leader for this project. It's doubtful it would have worked without her. She had mastered the difficult art of being a member of the team as well as its manager. Nicole had excellent management skills, always finding a way of balancing the individual needs of her team with the greater needs of the project. These weren't graduate students who needed endless direction; they were experienced scientists. Like the veteran manager of a sports team, she managed to create an environment that allowed her three research assistants to utilize their skills. Everybody was actively involved every day.

An essential aspect of the Japanese production model adapted by American industry is that every worker believes he or she has a personal stake in the final product. That's precisely what Nicole did, beginning with giving the project a real name, not just the normal scientific nomenclature. Calling it the Rick Project humanized it. It made it impossible for anyone to forget that this wasn't a typical research or development program, but a crash project to help a specific person.

From the very beginning of the project, Nicole did everything possible to alleviate the pressure. "Look," she told them at one of the early meetings, "we're only going to get a few shots at this. We don't really have enough time to do it right, so if Rick croaks, it won't be because we failed. It'll be because the chemo failed or a technician dropped a bag, or he gets too much radiation.

"There are endless variables in this project. We're not the only cog in this wheel. What we're doing is only the first step. It has to work for the transplant to be successful down the line, but even if we don't get every tumor cell, we're going to give him a better chance than he has right now."

Nicole also limited the hours they would work. When good scientists get involved in an exciting pursuit, time ceases to have meaning. The same people who so carefully calculate data to several decimal points easily lose contact with time. Minutes, hours, even days can ease by without much notice. In a situation like this one, in which a vast amount of work had to be done in a limited period, there existed a real danger of burnout. To avoid that, Nicole insisted that the team take off weekends, and made sure no one worked more than one or two nights in a row.

The organizational stages of the Rick Project had actually gone well. They had the antibodies and a prototype column. I don't think I ever saw this original device, but apparently it was something to behold. The actual Ceprate instrument was attractive in a futuristic sort of way, consisting of a bright white rectangular platform about three feet high, rising from an angled white console. Seven pinch valves, resembling empty film canisters, popped out of the vertical platform and several more emerged seemingly haphazardly from the console. Tubing ran between the valves into the column where the avidin-biotin cell selection occurred.

Although we referred to it as a column, in fact it was actually a round plastic cylinder that vaguely resembled an enclosed miniature gazebo or a tiny carousel. It was filled with the tiny, avidin-coated beads. Unseparated blood ran down through the tubing into the top of this column, and then slowly spread out over the beads. It was like watching a red tide gradually covering a beach. Eventually the remaining blood product flowed out the bottom of the column into a bag. Then the selected cells were released from the beads by gently rotating a magnetic stir bar, and diverted into a separate collection bag. It was a completely automated process. It was a beautiful medical instrument.

The initial Rick Project column didn't look like that, however. Like the breast-cancer prototype, it looked a lot like something that had been slapped together from whatever pieces they found lying around the lab.

Syringes were hanging off it at crazy angles. It looked as though it would fall apart the first time someone tried to use it. If the situation hadn't been so serious, this contraption would have been very funny. *Oh no,* Nicole thought when she first saw it, *we can't use that thing.*

But it worked. Each step took place at the point in the process at which it was supposed to take place. The device needed a lot of development—everything about it, from the antibody concentrations to the speed at which the fluid moved through it, had to be tested and retested until optimum values were obtained—but the design was adequate. And on this time line, adequate was more than acceptable.

Considering the vast number of variables, they made steady progress. But they couldn't get consistent results, which is the foundation on which experimental science is built. At times the system worked beautifully, and at other times it failed badly. The results of each run—each attempt to deplete tumor cells—were reported by the FACS as a number on a printout, or as glowing dots on a computer monitor. As the cells were sorted by this instrument, each time a fluorescent tumor cell passed through the counter, a dot would pop onto the screen. After each run, Stan, Sharon, and Kirsten would gather optimistically around this screen to watch as the results of this latest experiment were reported. They waited, they watched, and they hoped.

Invariably, little dots began popping onto the screen, each one representing a tumor cell that had fought through the system. The depletion system just wasn't working.

It was the ultimate scientific dilemma. They knew that at the end of the process was the pot of tumor-free stem cells; sometimes they got close enough to touch it. But for some unknown reason they just couldn't quite get there. Each step in the process had been tested successfully, individually they all worked, but together they didn't produce the desired result. At the end of some runs there was an abundance of almost tumor-free stem cells, yet other runs that reported similar results at each step ended up with a paucity of stem cells.

The loss of stem cells somewhere in the system was cataclysmic. If that problem couldn't be solved, we couldn't risk using the B-cell depletion system during my transplant. I'd still get my stem cells back, but with them would come a load of tumor cells.

When the team met each morning, they would analyze the previous day's experiments and decide how to move forward. A lot of science is theoretical; this was about as practical as possible. It was being done real-time; the ideas of the morning were the experiments of the afternoon. The team would sit in the conference room, crunching orange Chee-tos, analyzing results, agonizing about the problems, tossing ideas on the pile, sketching representations of this invisible world, designing new experiments, desperately trying to move the project forward. There were so many variables in this system, so many alleys to wander down. Was the source of the problem the antibodies? Were there too many or too few? Was the flow rate too fast or too slow? Maybe it was the configuration of the column. Or the blocking reagents that had to be added.

The normal way to identify the source of a problem is trial and error, trial and error, trial and error. Do it again, then again, and again, changing one variable each time. Good science requires great patience. But in this case they lacked the time to be patient.

The days began stretching into the evening. Everyone usually got to the lab by nine o'clock in the morning, no matter what time they'd left the previous night. Mornings were spent in meetings or preparing experiments. But the day really began when the blood arrived in the early afternoon. Then the team burst into action, and one of them would be in the lab until late that night or the early hours of the next day. By 9:00 A.M. they would be back and starting again.

The result of every experiment was a number. How many or how few. Centuries ago, experimental science began with the use of everyday objects to try to decipher the laws that governed the physical world. Balls were rolled down inclines, objects of different sizes were dropped from heights, candles were burned in jars. With the invention of the microscope, another world was discovered, the world in which billions of almost invisible creatures live in a puddle of water. Under the microscope, cells would divide and bacteria were visible, a drop of blood could be divided into its many components. But still it was a visual world, a world that could be seen.

Most of the work we did at CellPro was on a submicroscopic level, much too small to be seen even with great magnification. The Rick Project experiments were biochemical reactions: How much biotin is needed to bind with these antibodies? Does this antibody effectively bind to that molecule? How many tumor cells are remaining in this concentration? Chemicals, reagents and antibodies, biotin and avidin, were injected by a syringe into specialized bags where the experiment actually took place.

There wasn't much to see. There was no fizzing or foaming, the reactions didn't bubble or gurgle, there was no danger. Sometimes a liquid changed color. Results were determined by sophisticated laboratory equipment, or calculated by computer. The drama was all in the numbers.

On occasion I would meet with Nicole or her whole team in my office. They'd bring in their data and we'd discuss what had been achieved and where they were going. I tried as much as possible to act dispassionately, as if I were reviewing an ordinary project. I found myself asking a lot of questions. At some of those meetings, when the data was not very good, that was difficult. "This is interesting," I'd say as I leafed through pages of numbers. "Why did we lose so many cells on this run?"

Just as I'd tried to read the verdict on the faces of the jurors as they walked into the courtroom, I looked to the expressions on the faces of Nicole and her team as they came into my office, searching for a sign of success. Most of the time I didn't find it. Stan would force a smile, Sharon would be looking elsewhere, Kirsten might be shuffling papers. Nicole was direct. *It still isn't working,* her eyes would tell me. *This is a problem, Rick, and we don't know how to solve it.*

Everyone in that room had sat through hundreds of similar meetings, yet none of us had ever been to this place before. It was almost as if we were on a movie set, playing our roles, reciting our lines, but studiously avoiding the dragon standing right behind me.

I never offered any suggestions. What could I tell them? Hurry? They knew so much more than I did. The most I did was nod my head, thank them for their struggles, and try to be encouraging. But I knew the odd sight of me sitting in my office wearing my hat served as both a reminder and a plea. Work fast, and help me if you can.

The first breakthrough came after more than three weeks of frustration. Until that time the team had been following the breast-cancer model. It was a two-step process: first use the Ceprate system to select for stem cells, then run that purified suspension through a second, smaller column to pick out the remaining tumor cells. But no matter what adjustments they made to the system, after the second step they were ending up with either too many tumor cells or too few stem cells.

Sometimes it's necessary to step back in order to move forward. Frustration leads to innovation. And that's what Nicole did. Like treasure hunters convinced the mother lode lay just beyond their reach, certain they would find it if they just kept digging, the team focused their experiments on optimizing the conditions and the process. But Nicole wondered if they were simply digging in the wrong place. She began looking at the obvious things, the steps so big that they can easily be overlooked, the fundamentals. One morning in early May, as the team sat around the table in the meeting room, Nicole pleaded for new ideas. "Come on, guys," she asked, "give me some input. Where are we going with this thing?"

They sat in that room for more than two hours, tossing ideas on the table for discussion. Finally Nicole asked, "What do you think would happen if we reversed the process? Suppose we did the depletion first, then the selection?"

In retrospect, this was an obvious thing to try. But only in retrospect. In real time it was not a natural thing to do at all. Even after Nicole suggested it, no one snapped his fingers and exclaimed "Eureka!" In fact, scientifically there was no reason why reversing the order should have made a significant difference. But it did. No one understood why; it just did. When the team depleted B-cells first, then selected stem cells, they ended up with a bagful of stem cells and only a few B-cells. Now they had a reliable measuring stick by which to gauge their progress. There were still too many tumor cells left in the mix, but this put them back on the right path.

I got most of my information about how the team was proceeding from Joe Tarnowski. Joe could justifiably wander through the lab because it was his job to know everything that was going on there. But then he would quietly report to me. I was at my desk late one afternoon when he knocked twice on my open door and leaned inside. "Hey, Rick," he said casually, "I thought you might want to know that it looks like they're making some progress with this reversal thing. The results seem to be a lot more consistent, they're getting good B-cell depletion and good stem cell yields. It looks like it might work."

"Hey, that's great," I said. "Thanks, buddy." Then I pretended to go back to work. But my heart was pumping much too fast for me really to concentrate. Instead I leaned way back in my chair and gave a loud, silent cheer.

We'd taken the first step.

8 OFTEN, LATE INTO THE NIGHT, THE ONLY PERSON
left working at CellPro was Sharon or Kirsten or Stan, waiting
patiently in the laboratory for an experiment to end. It was a
lonely place to be. Usually, starting about five o'clock in the evening, the building slowly emptied. The lively sounds of a growing company doing business were gradually replaced by an eerie electronic silence.
Rather than voices and ringing telephones, people walking through the
halls or loudly completing a crossword puzzle in one of the common
rooms—the noises of everyday life—the night sounds were the soft buzzes
and hums of computers and small motors powering laboratory equipment.

For a time, my disease affected their lives far more than it did mine.
While they worked into the night, and worried, I was at home with Patty
and Ben or on a brief business trip, trying desperately to maintain a normal
life. With the exception of two enlarged lymph nodes, I had suffered no
symptoms. I rarely thought about the battle being fought inside my body, a
battle I was losing just a little more every day as my diseased cells continued
to multiply unabated, silently laying siege to my immune system. Only
when I had to put on my toupee in the morning, or wear a hat during the
day, or when I glanced in the mirror, did I remember that I was dying.

Patty and I rarely spoke about the disease, except as it affected our
daily schedule. Just as Patty had once planned her life around the boys'
needs—who had to be delivered to what activity at what time—so now it
was my needs that had to be met. But even in our most private moments,
neither of us dared show a sign of weakness. Our defense mechanism was
to deal with this thing as if it were just a part of life, rather than a harbinger of death.

Had I known what was going on in the lab, I probably would have
been less sanguine. The first step the team had taken had led directly to a

154

wall. As often happens in experimental science, the solution to one problem had created an array of new ones. Many months earlier, Sharon had predicted that using avidin and biotin in two successive steps would cause a real chemical mess. Switching the order in which the steps were done, picking out the tumor cells first then running the solution through a second column to select stem cells, had provided greatly improved results. At the end of this first step the team was ending up with an acceptable 0.2 percent tumor cells in the mix. Not perfect, but very good, and much better than they'd been able to do before.

The problem was with the second step. In this phase, biotinated antibodies were attached to stem cells and run through a column of avidin-coated beads. Unfortunately, the B-cells that escaped the trap laid for them in the first step were still coated with biotinated antibodies. So both the biotinated stem cells and the biotinated tumor cells locked on to the avidin-coated beads. As it turned out, we were actually reselecting for tumor cells. By the end of the second step, the percentage of tumor cells in the final collection had increased tenfold, to 2 percent.

The team tried adding a second purging column in the first phase; they stacked one on top of the other and referred to it as a condo system. Theoretically, that gave them twice the opportunity to pick out tumor cells. But practically it had only a minor benefit, not enough to make any real difference. Like enemy soldiers infiltrating an army, the biotinated antibody-coated B-cells continued to slip through defensive lines.

A lot of biochemistry takes place on paper. Just as children use stick figures to represent people, scientists use rough drawings to represent chemical reactions. In nature, molecules and antibodies aren't little balls with rods running out of them, but these representations are very useful. A scientist plans his experiments on paper using these symbols, in the same way that a football coach diagrams plays with X's and O's. What might look like gobbledygook to most people is in fact a game plan for scientists. It's a mathematical problem that needs a solution. In these daily meetings, Stan and Sharon and Kirsten filled endless pages with diagrams, trying to determine what would happen if . . . suppose we did . . . how can we . . . maybe if we add some . . . Avidin, for example, was represented as a ball emitting four spokes, each spoke a potential landing site for biotin. As long as any one of those landing sites remained open, biotinated antibody-bound tumor cells could land there. The alternative was to give the biotin another place to land. As a football coach planning defensive strategy might move a tackle to an open spot in his defensive line, the team realized they needed to offer biotin a place to land before reaching the column. The answer appeared to be what is called a *blocking reagent.*

Part of the appeal of the Ceprate system to transplanters was its simplicity. You put in blood product at the top, and concentrated stem cells came out the bottom. The fewer additional steps necessary, the easier and safer it was to use. No one on the team wanted to use blocking reagents in the process; it would cause tremendous complications. It might even cause the entire process to fail. For the Rick Project, for a one-time use in what was essentially an emergency situation, it would be acceptable. But as part of a marketable product it would make a simple procedure very difficult. And if it failed, if a technician made even a small mistake, it would fail catastrophically.

The team had no other viable options. A reagent is simply a fancy name for any substance involved in a chemical reaction. It's part of the recipe. It's the spice that makes possible the unique taste, it's the cornstarch in custard, it's the gelatin that makes Jell-O wobble. It might be a liquid, a powder, or even a crystal. In this situation the team used two reagents, solutions of avidin and biotin. The theory was that these reagents would block all the available sites on the biotinated B-cell antibodies before the second step, thus preventing the B-cells from binding to the avidin on the stem cell selection column.

On paper, it worked very well. On paper, all those spokes got filled up. The question was whether it would work in the lab. Finding the answer to that question would be a laborious task. It wasn't as simple as putting "approximately two cups" of detergent in with the laundry. The assumption was that this would work only if the correct concentrations of the reagent were added; the difficulty lay in determining precisely those correct concentrations.

The meeting at which the decision was made to use a blocking agent was two big bags of Chee-tos long. At best, it was a makeshift answer to a pressing problem. No one on the team had any real experience in using blocking reagents, so the only real way to determine how much to use was to experiment. And experiment. Hundreds of hours of work would have to be condensed into several weeks. But no one complained. By this point, I think, this project had become more than a job. And it wasn't so much their dedication to me as much as it was their lifelong curiosity about science. This was an intellectual puzzle that needed to be solved.

Rather than fitting the Rick Project into their lives, they molded their lives around the needs of the project. They spent more time in the lab than at home, more time with each other than with their own families. As Stan once commented, "We became each other's shadows." Although it was my life that was at stake, their work had a great impact on their own lives. For each of them it had a very different meaning.

Stan gradually emerged from his depression. His responsibilities to the rest of the team got him up off the couch at home and reminded him, he told me, "How precious life is, and how often we take it for granted."

Shortly before the project began, Stan had joined the controversial religious association of men called the Promise Keepers. Stan had always been a relatively religious person, and his membership in this group had caused him to reexamine his relationship with Christ as well as his fellow man. And he decided he was falling short. The Rick Project offered him the opportunity to give of himself to someone else who needed help. "The way I looked at it," he explained, "was that God put us on this world to be at the right place at the right time, and that He gave me the skills that I was supposed to use. It just so happened I was in the place I needed to be, with the skills I needed to have, and I was working with the right people."

It took him a little while to trust Sharon and Kirsten enough to confide in them. Sharon sort of had to drag the truth out of him. After a particularly difficult afternoon she finally demanded, "What's going on, Stan? What's the matter? You've been a real pain since the day we started!"

Reluctantly, he finally revealed that his engagement had ended badly and he felt lost. In addition, his teenaged daughter from his first marriage had come to live with him, and they were not getting along very well. After keeping all this sadness inside for months, talking about it seemed to liberate him. The curt attitude he had displayed gradually disappeared. "I began to understand," he said, "that instead of dwelling on this failed relationship, there had to be another relationship that God wanted me to be in rather than this one."

Some scientists like to work to a musical accompaniment, while others savor silence. If there is music played in a lab, it's generally benign, chosen specifically because it won't offend anyone. There wasn't too much music played at CellPro, although several people did keep small radios in their work spaces. Stan had a radio, which he played softly and kept tuned to a Christian music station. If there was a soundtrack to the Rick Project, it was this religious music.

Stan had also started a small prayer group at CellPro that met twice each month. Among the reasons Stan started this prayer group was to offer an additional type of support to me. I may not be a very religious person, but I did find comfort in those prayers. And from time to time I might have even closed my eyes and offered my own little silent prayer.

Sharon had been in pretty bad shape emotionally when the project started. Several of the people she had been working with on the fetal-cell project had been laid off. Sharon's job title was RA-4, meaning she was a top-level research assistant. But she meant a lot more to CellPro than her job title indicated. She was an excellent scientist, the person other RAs came to for advice or backup—and for too long the person to whom she reported had her performing utterly boring routine tasks. She had been ready to quit. Initially she had been reluctant to join the project team, and Nicole had really had to persuade her to do it. "I really need you on this one," she'd told her. "You have to be my expert on this stuff. You have to rise to the occasion on this lymphoma thing."

And then she'd sent her into the game to lead the team to the winning touchdown.

For Sharon, probably more than for Stan or Kirsten, the nights in the lab were the most difficult. Some scientists love that feeling, that solitude. But not Sharon. The janitor, a large man, would start cleaning in the early evening. For a while his presence offered Sharon a sense of protection. But when he finished about eight-thirty and left the building, she felt more alone than ever.

There was something disconcerting about being all alone in a large, brightly lit building. While the lab in which she worked had many windows, she looked out into a darkness broken by only a few lights in other buildings. Usually she was so busy that she was able to lose herself in the experiments. The anxiety disappeared.

But every once in a while the night silence would be punctuated by an unexpected sound that would catch her attention. Was that a chair sliding? Did something fall? Had someone else stayed late? For a few minutes

Sharon's heart would start beating just a little faster, then calm down as she went back to work.

A casualty of these late nights was Sharon's relationship with her twelve-year-old son, Jason. With her husband spending so much time on the road, during the week Sharon often became a single parent. Normally she was there to drive Jason to his activities or to play baseball with him, but her responsibilities on the Rick Project robbed her of that time. Jason missed her. Several times during the day he'd call or page her to ask her when she was coming home, or to ask a routine question—but what he really wanted, she knew, was her attention. And every once in a while he'd complain, "I never see you anymore, Mom."

Sometimes, when it was her turn to work at night and she couldn't find someone to stay with him, Sharon would bring Jason to the lab with her. She'd told him exactly what she was working on and why she had to be there so late at night. And as she worked on her experiments, Jason would sit quietly at one of the lab tables, doing his homework.

But given the responsibility of too much work and too little time in which to get it done, Sharon responded to the challenge. This was the type of project she'd envisioned working on when she'd first decided to become a scientist. Maybe she wasn't going to cure cancer as she'd once dreamed, but potentially this was an important step in curing *some* cancers—and an opportunity to use her knowledge to make a difference in one person's life. The Rick Project was helping her rediscover the excitement of experimental science.

Nicole, too, put in the long hours. But often she spent those hours at home, reading pages of results, checking data, trying to understand what went right and why. By this time her son, Trevor, was almost eighteen months old and was enrolled for several hours each day in day care. Her husband, Frank, was spending the year on sabbatical, doing his own computer software design from the house.

Nicole and Frank work in very different worlds, but they each speak the other's language. At night, after Trevor was asleep, Nicole would matter-of-factly discuss the Rick Project with Frank. From day to day, Frank rarely noticed any change in her temperament. He couldn't tell from her mood whether the experiments done that day had been successful or had failed miserably. Nicole was a steady hand at the helm and was not about to change course to pick up every promising breeze.

For Frank, and for Sharon's husband Jerry, and for Kirsten's husband

James, this was simply a crash project that would end in a few months, enabling their lives to return to normal. Just part of the job. Maybe there was a lot more pressure involved, and certainly the challenge made it interesting, but still it required only a temporary adjustment.

Like me, they were mostly cheerleaders. For the first six weeks there was little I could do to help—except stay out of their way. My involvement would finally begin in mid-June, the first time one of the two rooms at the University of Washington Hospital in which Ollie's patients could be safely isolated after receiving high-dose radiation treatment would be available. In late May, Nicole and Sharon, Cindy Jacobs, Monica Krieger, and I met with Ollie and Scott Rowley to lay out a realistic timetable. We met in a large classroom in the teaching wing of the hospital, and sat in a large circle as we laid out the timetable for my transplant.

It was Ollie who finally asked Nicole, "How far away are you from being ready?"

Nicole glanced down at the note she was holding with two hands. "We need two more weeks, at least." The experiments using blocking reagents had been relatively successful, she reported. They still didn't have all the concentrations quite right, they had not yet been able to deplete every single tumor cell from their model system, but it was clear they were making steady progress.

"All right," Ollie said professionally, "how's June 17?"

June 17. For Nicole and Sharon and Kirsten and Stan, that would be the day their work would end. For them, it would be the last day of the Rick Project. For me, if everything went correctly, that would be the day my part would begin.

In early June I would receive chemotherapy and G-CSF, a growth factor, to mobilize my stem cells. When my marrow was pumping them out into my system by the millions, they would be collected by apheresis in two consecutive daily sessions. The result would be two small bags of white blood cells. Scott Rowley's team, under the direction of Sharon and Kirsten—Stan was committed to be in Colorado, training technicians to use the breast-cancer-purging prototype—would do the actual tumor-cell depletion and stem-cell purification.

They would have only one attempt to get rid of all the tumor cells. The second bag had to be saved in case something went wrong during the process. Ollie had to have enough stem cells to put back in my body, even if those cells were still polluted with tumor cells. Without that infusion, with luck I would live two weeks.

After being processed, my cells would be frozen in liquid nitrogen and sit untouched in the freezer for almost two months. But there would be no going back, no second chances. Whatever product was in that bag could not be altered in any way. That was my future in there.

In July I would enter UW to receive high-dose radiation. I would have to remain in total isolation until I was no longer radioactive. I had been told that while I'd probably be uncomfortable during this period, I would be able to function. I planned to take several books and a computer into isolation with me. There was a telephone in the room, and I expected to stay in close contact with the office. Ollie had told me that everybody tolerated the radiation treatment differently, but I knew myself very well. I was in excellent shape. It wouldn't be easy, but I was confident I would find ways to use that time profitably.

After completing the radiation treatment, I would receive the big blast of chemotherapy, enough to kill all remaining blood—and, we hoped, tumor—cells in my body. These drugs didn't discriminate between healthy cells and tumor cells, they killed them all. I had been through several cycles of chemo and had tolerated it reasonably well. But this time I was to receive as much as ten times the amount of drugs I'd gotten. Although I would also receive drugs that abated the worst effects of the chemo, I'd been warned that I would be extremely nauseated for several days.

The science of stem cell transplants had begun a half-century earlier, with the bombing of Hiroshima. I was going to receive a dose of radiation and chemotherapy as massive as many of the victims received that day. This was another kind of war. One day after I'd completed chemo, my purified stem cells would be taken out of the freezer, thawed, and put back into my body.

Ollie referred to that day as my second birthday. Through some miracle not yet understood, those stem cells would find their way into my bone marrow and engraft. Slowly they would begin multiplying. For several days there would be little evidence that a new immune system had taken root. This was the period about which the FDA was so concerned, the time during which I would be most susceptible to life-threatening infection, fevers, and uncontrolled bleeding. It could take as long as two weeks for my implanted immune system to begin functioning. And when it did, I would walk out of the hospital cancer-free, with the expectation that I would live a long and very happy life, hopefully.

Professionally, I prepared for my transplant as if I were going on a long business trip. I made sure Larry Culver, who would be running the company in my absence, had all the materials he might need. I cleaned my desk of pending business. And on May 30, I sent an e-mail to everyone at CellPro:

> *As you know, I have been undergoing treatment for non-Hodgkins lymphoma for the past four months. A recent reassessment of my prognosis has led my physicians and me to the decision that a new course of therapy will produce more positive results.*
>
> *We are in the process of finalizing the details that would permit me to be treated on a new protocol using a radio-labeled monoclonal antibody treatment that will be followed by a Peripheral Blood Stem Cell (PBSC) Transplant. I now anticipate that the stem-cell mobilization and harvesting process will begin within the next two weeks. My circumstance is giving me the unique opportunity to be the first patient treated using our prototype lymphoma purging system. A number of you have been making extraordinary efforts to get things ready for the procedure, and I am very grateful for your efforts on my behalf. . . .*

Mentally, I also felt prepared for this. I knew almost as much about what was going to happen to me as did the attending medical staff. More, even, in the area of stem-cell selection. I knew this was going to be an ordeal, and I tried to rid my mind of anything that would detract from my ability to concentrate on this and only this. I told myself over and over that this was going to be slightly more than a month out of my life, and it was a risk that I had to take because it was my only chance for a cure. If I was going to come out the other side, this was a crucifixion I would have to endure.

Oddly, at least a part of me welcomed this respite from the front lines of business. It seemed as if I had been in combat for CellPro for so long. The battle with Baxter had raged for years; the Corange deal had been exhilarating, then terribly depressing; the FDA approval process had been difficult and draining; the trial had been exhausting. In the past few years I'd spent as many thousands of hours with lawyers as I had with our scientists, doctors, and sales and marketing people. And I was beginning to hear disturbing rumors that Judge McKelvie was dissatisfied with the jury verdict. I was tired. I needed a rest, even an enforced retreat such as this one. So, in a sense, I looked at this process as a welcome escape.

I was ready for the transplant and I wanted to get on with it.

I tried to exude confidence for my family as I put my life in order. But I never lied to myself about my chances. I had spent my adult life in a world that exists on statistics. I'd seen all the numbers about mantle-cell lymphoma. I knew the survival rates and the mortality rates better than I knew my social security number. While I had complete confidence in Ollie Press and his staff, and I believed totally in the Ceprate system and the purging system, I also knew that for some diseases science has no cure.

I updated my will. Patty and I discussed it, as we discussed everything, and it was a simple thing to do. Our lawyer did it for us. We treated it as a normal part of life rather than as preparation for death. We realized we hadn't updated it since leaving California several years earlier. Washington State's laws were quite different from California's. It was time to do it, just like cleaning out the garage.

Both of my boys had steadfastly refused to accept that I was seriously ill. They treated my lymphoma like a temporary inconvenience. I tried to be as open as possible with them, but they didn't want to know the details, or the odds. After we told them, I don't remember speaking about it again until I began the transplant procedure. They had decided I was going to survive, and made no concessions to my illness. This was simply another episode in the life of the Murdock family: move to Brussels, join a new company, cheat death. Jamie was spending the semester in Venice, so our relationship consisted of the usual hurried phone calls and the occasional letter. But Ben was home, and almost every weekend he'd insist, "Let's go, Dad, let's take the boat out and go sailing."

Patty would usually object: I shouldn't be exerting myself. I couldn't afford to catch a cold. I was too weak. She worried about me constantly. And while she worried, Ben and I went sailing. It was the best possible thing I could have done, the best therapy in the world. The alternative would have been to sit around the house worrying about my future, driving everybody crazy. Almost every weekend Ben and I, sometimes with his friend Jason Dunn, would sail up Puget Sound into the wind.

There were some chores on the boat I could no longer handle. It takes two people to lift the dinghy on and off the boat, for example, and I had some difficulty holding up my end. Ben never said a word about it, but just put a little more weight on his own shoulders, as if we'd always done it that way.

As soon as I got out on the water, I was able to forget that I was sick. Hospitals and tests seemed part of another life. Only once did Ben and I speak frankly about my condition. I just felt it was time. We were alone on the boat, sitting in the cabin, sipping champagne. "You know, buddy," I said, "with this transplant there's a real good chance everything's going to be okay. But there's still the possibility it isn't going to work out."

Ben's always been a curious kid, but this night he asked no questions. As the boat gently rocked, I explained the transplant procedure to him. I didn't have to tell him what I expected of him, or that the most important thing he could do was be supportive of his mother. I said nothing like that. You raise a child as best you can, and then you watch and hope as he or she matures into a unique person. I love Ben and Jamie as fiercely as any father could love his children, but I was fortunate in that I also liked the people that they had become. Unlike me and my own father, my boys and I had become close friends. I had absolutely no doubt that whatever needed to be done, Ben and Jamie would be there to do it.

One Friday night, a few weeks before the procedure was to begin, Ben, Jason, and I sailed north into Saturday, sailed through the day to a place we loved called Marrowstone Island. Between Marrowstone and Indian Island lies an extraordinary inland bay. Getting into the bay is tricky, and you have to catch the tides just right because the entrance is so shallow. But because it is a difficult entry, the bay remains pristine. A little spit of land called Bishop Point reaches into the bay. Jamie and I had discovered it one weekend two years earlier, and had christened it Murdock Point. We've been there many times, but we've never seen another boat there. It was Ben's favorite place in the world, and just maybe it was mine, too.

We dropped anchor late in the afternoon. There wasn't a cloud in the sky. We set up our barbecue on the back of the boat, and toasted the sunset. That night was as clear as any night I have ever seen on the water. We were far enough from the lights of civilization not to be disturbed. So we lay in the cockpit listening to music while we watched the passage of the stars through the night. We tracked the occasional satellite and solved all the world's problems. The bay lay mirror-still, reflecting the great timbers reaching out toward us from Indian Island. It was as if this night had been set before me as a gift in preparation for what was about to happen.

At about one-thirty that morning I started to go below. The long day's sail had taken its toll on my energy, but as I stood up, I noticed a glow behind a nearby hill. As I watched, a late-night moon appeared. The moon was three-quarters full, a bright orange waning moon, and for a time it seemed to rest on the hilltop before resuming its climb into the night. It was awesomely beautiful, so I sat down again. Eventually this moon laid a silver path across the water to our boat. We sat in the cockpit till almost 4:00 A.M., completely entranced by this display. I had rarely been as calm as I was that night. It was a good place to begin.

▷ I had been warned what to expect by a dynamo named Sherri Bush, officially the radio-labeled monoclonal antibody nurse coordinator, but essentially the Ginger Rogers to Ollie Press's Fred Astaire. Sherri is a small woman—like Patty, she was barely over five feet tall—with long, dark hair and eyes eternally flashing with excitement. Growing up, she had worked in her family's Italian bakery in upstate New York. She was the family expert on cake icing, and proclaimed proudly that she could produce a cake any color a bride desired. Referring to Sherri Bush as a nurse is like calling Pavarotti a singer; she was a virtuoso in the difficult field of caring for human beings.

Unlike her wedding cakes, Sherri put no pretty icing on the truth. Patty and I had sat in her office one day while she shattered any illusions I might have had that this might not be so bad. She wanted to make sure I knew that it *was* going to be that bad. "Going through a transplant is the hardest thing you'll ever do in your life," she'd said flatly. "It's a very personal journey, and you have to take it by yourself. It's like being in a war and discovering you're the only person fighting for your country, the only one, while the enemy, cancer, has a whole army and attacks relentlessly. Whenever you let your guard down even for a minute, the enemy is going to attack, and the attack is going to be brutal.

"The only people who can help at all are your family, and all they can give you is moral support. When it really gets tough, they can make sure that you know the reason you're continuing to fight. This is something you've never done before, and, God willing, you'll never do again. And I don't care how well prepared you think you are, you're not quite prepared for what a transplant does to you."

Sherri's job through the entire transplant procedure was to maintain the patient's contact with reality. She ran the high-dose radiation program in a way that allowed Ollie and his assistant, Dr. Steve Liu, to do their work to the best of their abilities. She worked with patients from the admitting room through the end of their transplant, taking care of everything from filling out insurance forms to actually administering the antibodies. In addition to being the patient's liaison with the medical staff, she also served as the patient's advocate. "Sometimes I'm a nurse," she explained to us, "sometimes I'm the cleaning lady, sometimes I'm the taxicab, and sometimes I'm Ollie when he's not here." Mostly, Patty and I found, she was a friend.

At the very beginning of my transplant, Ollie made it clear that Sherri was running the protocol. Then Sherri bluntly laid out the rules. "I am the tiny dictator. I don't ask you to do anything you don't need to do. The most important thing you can do is communicate with me. I don't want any surprises, ever," she warned me. "I don't want to be finding out Monday morning that you've had a fever and a huge infection the whole weekend."

She handed me a piece of paper on which was written her home phone number. "You call me whenever you need to. If it's three in the morning and you have a problem, I need to know, because I don't like surprises. If Ollie knows something I don't, then I haven't done my job."

When she'd finished, I almost felt like saluting. She gave us a calendar on which she'd circled several important dates, the days on which the side effects would be the worst, the days I would most need support. "This is the time to mobilize the troops," she told Patty.

I had listened carefully to every word she said. And yet I still believed that my experience would be different. Maybe it would be tough, but certainly not as bad as she forecast. I had been through what seemed like endless tests in preparation for the high-dose chemo and the transplant. Ollie's staff had examined every one of my internal organs several times. They had given me a test dose, then taken sonograms and gamma-scans to determine how much radiation I could tolerate. For weeks I'd lived with marks on my body so that measurements could be taken in precisely the same spot each time, marks that Patty had to redraw with a special pencil after every shower. This thing was a monster constantly looking over my shoulder. Whatever the effects, I felt certain I was ready for them. I just wanted to get it over with. I wanted my life back.

As I checked into the hospital at about 6:00 A.M. the next day, the reality hit me. It had been more than thirty years since I'd been a patient in a hospital. While I was in high school, I'd gone to the beach with my church group. We'd been caught in a savage riptide, and there had been some panic. I'd helped pull several people out of the ocean. But one girl had drowned, a girl I knew. It was a terrible thing, the closest I'd been to death. They'd put me in the hospital for observation, but I suspect I was actually in shock.

During my career, I had spent a great amount of time in hospitals. But at the end of every day I went home. Not this time. This time I was there to stay. I was handing over control of my life to other people.

The Hickman catheter was to be inserted under local anesthesia in what is known as the cath lab. The preparation was done by a male nurse. I remember he had a very strange haircut and a huge ring in his ear. "Okay," he said, "I'm gonna get an IV line going here so we can start the procedure."

He took my arm, held it firmly, and tried to punch a needle into my vein. He missed once, twice . . . "Sorry," he said nervously.

I'd been a patient for less than half an hour, and thus far it was not going very well at all.

As always, Patty was there with me. We hadn't had time for breakfast, and as she watched this nurse making a bloody mess of my arm, she turned pale and leaned against the wall. During her own career, Patty had done similar insertions hundreds of times without having the slightest reaction. But watching this nurse unsuccessfully searching for a vein proved too much for her. There is a scene in the movie *Fletch* in which Chevy Chase, in the title role, has snuck into a hospital and, to escape detection has put on doctor's surgical greens and a mask. As he passes a room, a doctor performing an autopsy with great glee asks him, "Hey, come over here and give me a hand with this." When this doctor cuts open the corpse, the escaping odor pushes him back, and he says, "Oh boy, you never get used to that smell, do you?" Fletch, of course, is staggered by it. Then the doctor says happily, "Have you ever seen a spleen that big?"

To which Fletch replies, "No, no, not since breakfast." And then he just keels over backwards. It was a very funny scene from one of our favorite movies.

The nurse finally located my vein on his third try. He hooked up the IV. Later, Patty told me that as she watched she couldn't help thinking of

Chevy Chase. Then she said to me quite seriously, "I hadn't seen a spleen like that since breakfast." I started laughing hysterically. It was a huge release.

The anesthetic knocked me out. I didn't even feel the catheter being inserted. I was still only semiconscious when they wheeled me upstairs to begin mobilization chemo. But when the anesthetic wore off, the pain hit me big time. In all the months since I'd been diagnosed, this was the first time I felt any real pain. My chest felt as if it had been ripped open. I was desperate for a painkiller. I had to make that pain go away.

Finally, my participation in this experiment had begun. I don't really remember receiving the chemo that would stir up my stem cells. They gave me a massive dose of VP-16, which blasted the hell out of my bone marrow. It just rocked me. Within hours I was more nauseated than ever before in my life. Once, I'd crewed a forty-two-foot sailboat out of the Virgin Islands, bound for Miami. It was the boat of a friend of mine, a former Director of Biologics at the FDA. There were four of us on board, and a few hours out of St. Thomas, the wind hit us. The lights of Puerto Rico had been visible about four miles off our port side, but within seconds it was pitch black. Suddenly I felt just a puff of rain. And then, before we could get our sails down, we were caught in a late tropical storm. The storm was right on top of us; we heard the thunder crackling before it roared. Flashes of lightning turned the night sky an eerie blue. At best, our visibility was only a few feet. It was raining so hard that the drops felt like splinters ripping into my face. I was at the helm, but I had to turn my back to the weather; every minute I'd face forward for a few seconds to try to keep us on course. At times I'm sure the wind reached forty knots. And the waves were pounding us. I tried to sail a compass course, but the seas toyed with us, picking us off the top of a wave and slamming us down into a deep trough. Rarely in my life have I been seasick. But on this trip I never got my sea legs. The ocean turned me inside out. I felt terrible. My head was pounding, I was terribly nauseated, I couldn't focus at all. Eventually we turned back to the Virgin Islands.

Yet compared to this round of chemo, that was nothing. My reaction really shook me. I thought, *Maybe Sherri's right, maybe I'm not quite as prepared for this as I believed.*

The chemo did its job. My immune system was staggered. My white-cell count dropped precipitously. I lay in a hospital bed for three days with barely enough strength to get up. Patty stayed with me until visiting hours ended. When we returned to the States from Belgium, she had started painting again, mostly landscapes. Peaceful, colorful scenes. Time frozen in place. My hospital room was on the sixth floor. She referred to it as "a room with a view" because it overlooked a park on the grounds of the University of Washington, and in the distance it was possible to see the majestic Montlake Bridge. As I recovered from the chemo, she set up an easel by a window and painted the scene in watercolors.

Just as Sherri had predicted, after three days I had recovered sufficiently to go home, but she warned me to be extremely careful, as I was susceptible to infection. Something as normally harmless as a sneeze could prove very dangerous to me. I wasn't to leave the house, or spend time around people other than my family. The only time I was supposed to go outside was when I returned to the hospital every day to be evaluated. For the first few days I didn't feel like doing much else.

As my white-cell count continued to drop toward zero—exactly as it was supposed to—I began receiving transfusions of platelets and red cells. It would take five to seven days for my counts to hit bottom, and when that happened my bone marrow would kick into high gear and start pumping stem cells into my peripheral blood system. At that time, like a farmer bringing in his crop, Ollie would put me on the apheresis machine and harvest healthy stem cells.

But until that time I felt trapped inside the house. I was forced to relax, and I couldn't do it. As I began to feel better, doing nothing became extremely difficult for me. Coincidentally, this was the week Ben graduated from high school. I was feeling fine, but I was not allowed to attend his graduation. I wanted to go to that graduation more than Ben did. Several relatives came up to Seattle, including my mother, who was staying with us. Patty and I had had some pretty strong battles about that—including one memorable discussion the first time we'd met Sherri Bush. We'd been in the hospital for some tests, and while we were waiting, the subject had come up once again. As much as Patty loves my mother, she worried that I was so susceptible to infection that no one should be near me—even my mother. "You're not going to have any blood counts," she'd said. "I don't think anybody should be staying with us."

I tried to brush the whole thing off. "Oh, come on, Patricia," I'd responded, "you're making a big deal out of this. I'm gonna be okay."

Just at that moment, Sherri Bush swept into the room. Patty immediately tried to enlist her in her cause. "Richard's mother is planning to stay with us for our son's graduation, but he's not going to have any counts," she said. "You think that's a good idea?"

Sherri didn't even hesitate. She looked at both of us and shrugged. "Mothers can do whatever they want. They're mothers. I'll tell you, when my son graduates, I'm certainly going to do what *I* want!"

So my mother had come to stay with us. And all of them, Ben and Patty and my mother, had left the house on graduation morning in a great burst of excitement. The silence they left behind was deafening. The house had never been bigger or more empty. I've spent days by myself in the woods, I've sailed alone many times, but I have never been more lonely than I was those few hours.

Patty videotaped the ceremony for me, but I've never watched it. We had been planning a family celebration, but had canceled it when my mobilization was scheduled. Several of the kids in the neighborhood had graduated with Ben, and one of our friends had invited everyone to their house for a barbecue. In the afternoon everyone went over there, and again I was alone. I took it about as long as I could, but eventually my stubborn nature took charge. I felt fine, so I decided to go to the party. My son was graduating from high school, and I wanted to share that with him. So I just walked over the hill into my neighbor's backyard, thinking, *I'm ready to have some fun.*

Patty was stunned and not particularly happy to see me. I seem to remember hearing the word "crazy." "You shouldn't be here, Richard," she said, "I don't want you getting an infection."

"I'm fine," I insisted, "nothing's going to happen." By the time we got home that night, I was absolutely exhausted, but exhilarated. I'd regained just a little control over my life. And when I woke up the next morning I was absolutely fine. At the hospital, tests confirmed that my bone marrow had finally started mass-producing stem cells. My white count was rising fast. It was harvest time.

Meanwhile, at CellPro the pressure on the Rick Project team had increased to the boiling point. The countdown to prime time had begun the day I went into the hospital. The purging system had to be ready to go when I mobilized. There had been no rehearsals, no practice runs. They had to hit a grand-slam home run their first time at bat.

Nicole remained cautiously optimistic, considering her feelings about the favor she'd been asked to do. Perhaps she was right. By traditional standards it was bad science, but it was still better than anything else in existence.

Incorporating the blocking reagents into the system had proved to be extremely difficult, requiring the expertise of every member of the team to find the delicate balance among all of the required elements. And getting it done once, for me, was not going to be enough. In addition to trying to save my life, they were also creating a product that we all hoped would be used many times to save many other lives. So the steps they incorporated into the system had to be simple and reliable enough for medical technologists to do on their own. The team wasn't just playing one game, they were creating the game board, and the instructions had to be written plainly enough to enable others to play the game safely.

Biochemistry is a vast, complicated puzzle. To those who understand it and love it, the world of molecules is endlessly challenging and frustrating and rewarding and entertaining—a world of drama and humor and mystery. In some ways it always can be reduced to the question that fascinated Sharon Adams and so many millions of others as children: *What would happen if I mixed these two things together?*

Every molecule or compound has known properties. It has a molecular weight, and it reacts with certain compounds to create new compounds, but will not react to others. The challenge is to manipulate those properties to your advantage. For some reason, reversing the two processes, doing the tumor-cell depletion first and the stem-cell selection second, provided the consistency of results that had been missing. But because both processes relied on the love affair between avidin and biotin, biotinated tumor cells were being reselected in the second phase. The team had to figure out how to prevent the biotinated tumor cells that had survived the first run through the column, the depletion phase, from being attracted to the avidin column in the second run, the selection phase.

The best way to block biotin is with avidin. Just like the space shuttle coming in for a smooth landing on a runway, when free avidin was added to the mix after the completion of the first step, the depletion phase, it landed on the remaining biotin antigen. It filled up all the landing sites. The biotin was blocked. But because avidin has four landing sites for biotin, some of those landing sites on the avidin that had just been added remained vacant. These were precisely the kinds of problems the team had tried to avoid by using magnetic beads for the second phase of the process.

During the second phase, when the antibody-covered stem cells were run through the system, instead of being attracted to the avidin on the column, where they would be collected, some of them were binding with the free avidin and flowing right through the system. As a result too few stem cells were collected to proceed with my transplant. The only way to prevent that was to add another reagent, more free biotin.

After I had my transplant, Patty wanted to understand better exactly what was happening. So she went to CellPro and asked Stan to show her how this system worked. An excellent teacher, Stan very slowly took her through the entire process, showing her everything that had to be added and explaining exactly what was taking place. Patty had been trained as a medical technologist. Although it had been many years since she'd actually worked in a lab, she still had a working knowledge of molecular chemistry and cell biology. In frustration, about halfway through Stan's explanation she asked him, "How long did it take you to really understand the whole process?"

Stan thought about it for a moment, then decided, "Probably about six months."

This is a very complicated series of reactions. Understanding it is difficult; working it out under this pressure was extraordinary.

If the reagents were added in proper concentrations, all the binding sites of the avidin and biotin would be blocked at the right times, enabling the tumor cells to be removed and the stem cells to be captured and collected.

It wasn't going to be elegant. There can be great beauty in chemistry, but not in this case. As functional chemistry, it was the equivalent of adding big chrome bumpers to a Lamborghini. But it made sense. If everything worked as expected, the result would be a tumor-free collection of stem cells.

If. A very big *if.* If everything worked as expected. And nature does have a habit of doing the unexpected. The key to the working of the whole system was the numbers. How much free avidin had to be added to block the biotin? How much free biotin was then needed to block the avidin? The numbers were critical. Just as a recipe can be completely destroyed by a single mistake in calculating the ingredients, this, too, depended completely on having enough, but not too much, of every reagent.

The only way to figure it out was the bedrock method of scientific experimentation, trial and error. There was no way to calculate accurately how many tumor cells would bind to the column, or how many would survive the purging process. Even in the most sophisticated laboratories, experimental science often begins with a well-educated guess. To paraphrase the commercial: Just do it—and then analyze the results and do it again, then again.

The team ran countless experiments, trying to determine the correct ratio between the numbers. It was the ratio that was important—how much free avidin to how much biotin, how much free biotin to how much free avidin—not the precise numbers, because the numbers would change. The first time we did this, the amount of reagent to be added would depend on how well I mobilized. But after my transplant was complete, I was determined that we would continue development of the lymphoma purging column until it was ready for FDA trials, no matter how long it took. Eventually this thing was going to work. And when it did, I wanted every other person facing what I was facing to have the same chance I was getting. Something good was going to come out of all this.

Eventually it would be approved and we would market it, and when we did, we had to make certain that every single time it was used, medical technologists knew precisely how much of each reagent to add to the solution.

In an ordinary biotech development program, dealing with this kind of problem is an inherent part of the process. We know it's going to come up, and we allot sufficient time to solve it. But obviously this wasn't an ordinary program. As Nicole had told Joe Tarnowski, a project like this would normally require nine months to a year before it was ready to begin clinical trials. The Rick Project team had only eight weeks to get it done. And even that time was running out. Less than two weeks was left when the team was finally ready to do a complete run with the model system. I had not followed their progress on a day-to-day basis. I knew they had been making progress, but I didn't know how close, or how far, they were from a working prototype. So I didn't know when they were going to rev up the system and see if it worked.

They were using donor blood spiked with fluorescent tumor cells in these experiments. The first time my blood would be used would be for keeps, as we used to say.

Early one afternoon in the sixth week they ran this blood, which contained billions of healthy cells as well as deadly tumor cells, through the B-cell depletion system for the first time. In a small sense it was like the Wright Brothers' first flying machine chugging down the launch ramp, trying to get airborne. The biotinated B-cell antibodies targeted B-cells and hugged tightly. When this mixture flowed past the avidin-coated column, the biotinated antibody was attracted to the avidin and locked on—capturing the B-cells. Then the column was removed, and with it went almost 99.9 percent, three logs, of tumor cells. In a sense it was like throwing out death. The purging system had worked exactly as designed.

The team was left with a bag containing mostly white cells, among them the desired stem cells—but hidden in there were still too many tumor cells. The team added the carefully calculated amount of free avidin to the mix and let it incubate. Although it looked as though absolutely nothing was happening inside that bag, in fact several billion molecules were playing musical chairs, searching desperately for an open place to land. It was Times Square on New Year's Eve.

After giving the reagent sufficient time to find its biological partners, the team added free biotin to the mix—and once again the dance of the molecules began. Only after this step was done did the second phase, the harvest of stem cells, begin. If everything worked as designed, all the remaining B-cells would flow right through the system into the waste bag, leaving only the stem cells attached to the column.

The entire procedure took slightly more than three hours. The end result was an invisible collection of cells. Literally, life in a plastic bag.

It was early evening when Sharon, Stan, and Kirsten gathered around the FACS machine. This was to be their scoreboard. The FACS was sensitive enough to detect one tumor cell in 5 million cells. If any tumor cells remained in the suspension, they would appear on the monitor as bright dots. As the FACS sorted the cells one by one, the team waited silently and watched, their eyes riveted on the computer monitor. The FACS hummed softly. Numbers flashed on the screen. No matter how many experiments a scientist has done, the excitement of discovering whether or not a particular experiment has worked never disappears completely. The immediate result is exhilaration, satisfaction, or frustration, but whatever it is, that feeling generally doesn't linger very long. The result of one experiment marks the beginning of the next experiment.

They watched and waited. "Looks pretty good," Sharon said softly. Years spent working in laboratories had taught her not to get overly excited about a single experiment. It was just one step, she knew; even a successful experiment meant that much more work had to be done. One experiment really didn't mean very much. You had to be able to get the same results over and over. But this time she couldn't help it. Her eyes remained riveted on the FACs, watching. . . .

"Wait," Kirsten cautioned, as if afraid saying even one word might change the result. "Wait."

They waited still longer. One by one the cells slipped through the FACS. No bright dots appeared on the screen. Finally the counting was done. As one, the three of them smiled, as if they had been expecting this result. The collection was clean. The blocking reagents had done the job.

Stan paused briefly, just to make certain the run was finished. It was almost too good to be true. He could barely believe it; they'd done it. "All right!" he cheered, raising his hand high into the air. Sharon slapped his open palm, then exchanged high fives with Kirsten. The purging system had worked. All those hours, the long nights, all those the bags of Chee-tos, they had all been worth it. They had a system that worked.

That night they closed the lab early and everybody went home, warm with the satisfaction of a job well done.

Once. The only way to prove the system really worked was to repeat the same experiment until the results could be predicted and duplicated every time. They all knew that wasn't going to be so easy. As Nicole had feared when the team committed itself to using blocking reagents, this was not a user-friendly system. There were too many places at which the slightest error would cause the whole process to fail. If the user mistakenly added the biotin before the avidin, the system would fail. If the user added too much or too little of the reagents, the system would fail. If the user added the antibody in the wrong order, the system would fail. If the flow rate was wrong, if the ratios were out of balance, if they didn't let the solution incubate long enough, if any one of many different parameters was wrong, the system would fail. But it had worked once.

Nicole didn't tell me that this initial experiment had been successful. She didn't want to give me any false hopes. "It's coming along," she told me. "We're making good progress. Looks like we'll be ready."

Joe Tarnowski elected not to tell me. "It's looking good, Rick," was the strongest commitment he would make. "They're doing a hell of a job."

The actual date of the transplant depended on my ability to mobilize. After leaving the hospital I stayed home for five days, wondering what was going on inside my body. With the exception of a little nausea and some weakness, I didn't feel very different from normal. Perhaps the most unusual feeling was being at home during the day on weekdays. I felt totally out of sync, as if I were living somebody else's life. Certainly since we'd moved to Seattle I'd never been home on a weekday. With Ben's graduation, things around the house were incredibly hectic. I felt a bit like an intruder, like the kid who stayed home from school. Everything was new to me. I didn't know what time the mailman delivered the mail, I hadn't met the neighborhood UPS man, I didn't know about the phone solicitors who called during the day.

I didn't know it was so quiet.

There was so much work to be done at CellPro, and I was trapped in the house. I read, I spoke to Larry Culver several times a day on the phone, and I cleaned up my desk. After two days my desk was cleaner than it had ever been. But I cleaned it again.

Occasionally, during this time, Larry and I would discuss the lawsuit. But it wasn't a major topic of conversation. Although we knew it would not be completely finished until Judge McKelvie signed off on the jury verdict, we weren't that concerned about it. Baxter, Hopkins, and BD had filed several motions with the court to overturn the verdict. That was normal; it's what the losing side does in just about every case.

I accepted other small changes in my life without complaint. The big event of each day for me was returning to the hospital to have my blood analyzed. For the first time since Patty and I had been together, she drove and I sat in the passenger seat. The view from the passenger seat was unsettling. I was used to driving, used to being in control. I turned out to be a bad front-seat driver. It was hard getting used to the new angles. Cars making turns looked as though they were about to hit us. I thought Patty was drifting too close to the center line. I found myself involuntarily pressing my foot down on an imaginary brake. The easiest thing to do, I finally discovered, was simply to close my eyes and try to sleep.

From the first day, Patty had been a bulldog. She had kept her eyes and her mind focused on doing whatever was necessary to get through this. Implementing, as she referred to it. She went with me to every meeting with my doctors and took copious notes, she took me to every test, she stayed with me in the hospital as long as possible. When I had to stay in the hospital overnight, she pretended I was simply away on a business trip. Gradually she started taking over some of my responsibilities at home. As she had when we'd first been married, or when I'd been away for a long time, she started paying the bills. The only thing she just couldn't do was the daily cleaning of my Hickman catheter.

The area where the catheter entered the skin and vein was basically an open wound, and my blood platelets wanted to seal it by forming a clot. I couldn't allow that to happen. Blood clots would clog up the catheter, making it impossible to use. To prevent that from happening, I had to follow an elaborate cleaning procedure every day, then inject heparin, a strong anticoagulant, into the tube. It took about twenty minutes and required changing bandages and alcohol and adding several reagents. One day, as Patty and I were in the bathroom lining up all these materials, she suddenly she got very lightheaded. She had to sit down. "Are you all right?" I asked.

She was really shaken. For someone who'd spent as much time in hospitals and clinics as she had, this procedure was nothing unusual. It wasn't even very bloody. The difference, she told me, was that this was family. She tried, but she just couldn't do it. "I'm sorry, Richard," she told me, "I can't do this. You're just gonna have to do it yourself."

"It's okay," I told her, "it's okay. Honest, it's no big deal." I reminded her that when I was in isolation I'd have to clean it myself anyway. But to Patty it *was* a big deal. A member of her family needed her, and she couldn't help. But that was about the only thing she didn't do.

As I waited at home for my immune system to mobilize, I finally received an e-mail from Nicole, reporting the progress of the depletion column:

Just a quick note to say hello and let you know what's been going on . . . all the materials have been produced and tested, and are ready to go. The newly purified antibodies still have a bit of endotoxin in them, but still far less than before. We're shipping columns, tube sets, and antibodies down to the Hutch. . . .

Sharon and Stan went to Scott Rowley's lab and trained his lead tech last week. By all accounts everything went smoothly and everyone feels comfortable working with the prototype and each other.

I talked to Scott Rowley today to go over last-minute details. . . . Since they are moving their lab on Wednesday the 19th (can you believe it?), we offered to run interference, lift boxes, and tote bales if necessary. If the move occurs before or during (God forbid) the processing, we'll need to get the Ceprate instrument recertified.

I hope you're feeling better and keeping your spirits up. Is there anything we can do for you? We're all thinking of you here, take good care of yourself. Best regards, Nicole.

On Monday morning, June 17, the Rick Project team began final preparations. The weeks had become days, the days were now counted in hours. Sharon and Kirsten packed up the prototype depletion columns and delivered the entire system, including the antibodies and reagents, to Rowley's lab at the Hutch. Although they had practically invented this system, they were not legally certified to operate it in a clinical situation. Stan and Sharon had trained Rowley's staff in its operation. Sharon and Kirsten would both be there for the actual processing of my blood. They were allowed to perform such basic lab work as doing cell counts and weighing blood bags, and they could offer advice and suggestions, but they were not permitted to touch anything. Stan was committed to a trip to Colorado, but he'd be waiting anxiously by a telephone to learn the results of the purging process.

As Sharon and Kirsten drove to the Hutch that morning, they felt a twinge of the emptiness a mother experiences when she brings her young child to school for the first time. With Stan, they had spent almost every waking minute of the past eight weeks creating this device, and now they were turning it over to strangers, to technicians who saw it only as another interesting medical device, people who knew absolutely nothing of the sweat and anger and pride that were part of its creation. When these people looked at it, they wouldn't see Nicole sitting up at home late at night poring over data, or Sharon tending her experiments while her son sat at a lab table, hunched over his homework. They wouldn't see Stan in there, finding renewed purpose in his life, or Kirsten dealing with the long separation from her husband.

But they drove away from the Hutch that day with the satisfaction of knowing they had done the best job possible under extraordinarily trying circumstances. Given the limited time and the resources, even getting to this point had been an incredible feat. If the purging system worked as well in the clinic as it had in the lab, I would receive tumor-free stem cells. If.

It was show time.

9 ON MONDAY MORNING, JUNE 17, 1996, I CHECKED into the apheresis unit at the Fred Hutchinson Cancer Center to begin the transplant process. There were several beds in the room, and in two of them were young children, kids between the ages of five and seven, suffering from leukemia. It was obvious they had already been through chemo, because they had lost their hair. They were hooked up to the apheresis instruments. One of them was crying. For a moment I thought Patty might start crying when she saw them. In a perfect world, little people aren't supposed to have to deal with big-people problems, but I really wasn't that surprised. In my mind I had become a patient, a member of a parallel society, and I knew from the years I'd spent in hospitals that in that world I was going to see a lot of unsettling things.

In fact, I was optimistic for them. Although leukemia was once almost universally fatal in children, the cure rates now are actually very high, as much as 80 percent. And I knew that at Emory University in Atlanta, Doctors Andy Yeager and Kent Holland were having real success using the Ceprate system to treat children with blood disorders.

Every patient responds to mobilization therapy differently. Some people just don't produce sufficient stem cells to go ahead with the transplant. I mobilized extremely well. My immune system had responded to the chemo attack by creating billions of stem cells in my bone marrow and pumping them into my peripheral bloodstream. The average patient produces about twenty stem cells per microliter. I was pumping out more than three hundred per microliter. When Ollie Press saw my counts, he said admiringly, "Wow, you've got great protoplasm."

The length of time it takes for stem cells to engraft after being transplanted, to find their way into bone marrow and start producing new blood cells, depends to some extent on the number of stem cells returned to the patient. More equals faster—up to a point. The number of cells necessary to go ahead with the transplant is based on the patient's weight; 5 million stem cells per kilogram is ideal. Although they can proceed with fewer cells, it probably will lengthen the time it takes for engraftment, and the patient will remain in danger longer.

The actual process of apheresis takes about three hours. The process was neither novel nor threatening to me. I'd been hooked up to many apheresis machines in my life. Two lines were connected to ports in my Hickman catheter, leaving my hands free. So, as my blood was pumped out of and into my body, I sat there in relative comfort, reviewing reports of clinical trials in progress and speaking with people in the office on a telephone. It was not quite so easy for the kids. Their parents tried to occupy them with games and toys, but these children were scared and restless.

As I sat there conducting business, all the blood in my system did three to five complete circuits through the instrument. The apheresis machine harvests that buffy coat, where the stem cells are found, and returns the rest of the blood components to the patient. Patients can easily survive the loss of billions of white blood cells without suffering any physical problem. Apheresis is a completely painless procedure, and when it's done the white-cell fraction has been collected in a sterile, clear polyvinyl chloride medical bag. Although the bag looks like it's filled with about a cup of a slightly opaque straw-colored fluid, it actually contains billions and billions of cells, all the different types of cells that constitute the immune system. It may also contain just a few million red cells, since the physical separation process is not nearly as precise as a chemical separation. And in my case it also contained several billion tumor cells.

Only a very small fraction of the billions of cells in that bag, usually less than 0.5 or 1 percent, are stem cells. The target cells. Once these cells are out of the body, they can survive for as long as one day without being touched, but then they'll die. In years past, standard therapy would have been to take this bag, just as it was collected, add the cryo-protectant chemical DMSO, and freeze it in liquid nitrogen. Once the cells are frozen, they can be safely stored for years. When the patient was ready for his transplant, the bag would be thawed and the entire collection of cells—including tumor cells—would be returned to the patient.

CellPro's reason for existence was to change that. Our B-cell purging device would eliminate most of the tumor cells. The positive stem-cell selection column would then pick out the stem cells. So, instead of receiving an entire bagful of cells, I would be given a suspension of cells smaller than the head of a pin. High purity stem cells.

The pouch containing the white-cell fraction was still warm when Sharon and Kirsten picked it up for delivery to Rowley's lab in the basement at the Hutch. As Patty and I walked out of the apheresis unit, we actually passed them in the corridor on their way. We greeted each other without stopping, and all sort of laughed cordially at the coincidence.

Patty and I went home. My day was done, but theirs was just beginning. Rowley's staff would do the actual work, but Sharon and Kirsten would be standing next to them throughout the entire procedure. They would supervise every step in the process. The day before, Sharon had taken one hundred dollars out of a cash machine to pay their expenses while at the Hutch. The first day she spent eighty dollars feeding Rowley's staff.

While we were still on our way home, in the lab biotinated B-cell antibodies were injected by syringe through a port in the bag. These were the B-cell—the tumor-cell—antibodies purchased from our French supplier, Diaclone. In our lab at CellPro, a technician named Joanne Cahn had chemically bonded the biotin to the antibodies. It was a simple benchtop operation. The downside was that our lab was not a sterile environment. Not even close. Doing it this way greatly increased the risk of contaminating the biotinated antibodies with potentially deadly bacteria that could produce endotoxins. In fact, the first lot Joanne prepared did contain unacceptably high levels of endotoxin, and had to be destroyed. The second lot also was contaminated, although not nearly as badly. Nicole told me, just as Joe had warned, "We're not sterilizing it, we're just filtering it, and you could get a really bad fever."

I told her the same thing I'd told Joe: "I'll take my chances. If I get an infection, they'll give me antibiotics."

The biotinated antibodies, which were as clear as spring water, were stored in a vial. A technician at the Hutch injected them into the bag of my white cells, then gently stirred the solution for about thirty minutes. The big dance was on again. These antibodies raced around searching for B-cells—healthy B-cells as well as tumor cells. The antibodies needed no preparation; embracing B-cells is the sole reason for their existence.

In real time, none of this was any more difficult than baking a cake. Until this point in the process, everything was basic lab work, biological bookkeeping. It had all been done before, countless times. But the next part of the process had never been done before with the life of a real patient at stake. It had required hundreds, perhaps thousands, of prior experiments to reach this point. It was time to attempt the first meaningful tumor-cell depletion.

The bag of cells was suspended from the top of the Ceprate instrument. Pliable tubing was inserted in the bag, and ran down through pinch valves—those round things that look like film canisters—into the depletion column. The column is that little plastic carousel. The flow rate from the bag to the column was regulated by a pump controlled by a computer program that also opened and closed the pinch valves. Both the depletion column and the selection column were filled with thousands of minuscule, avidin-coated plastic beads. The top and bottom of the column are covered by screens with openings about 40 microns wide. The beads inside the column are 300 microns, so they can't fit through those screens, while the cells are about 10 microns in diameter, so they flow through easily.

By the time my cells started flowing through the depletion device, I was at home, probably on the telephone, doing some business. And Patty was probably scolding me for not resting. I was completely disconnected from the process. But I do know that we didn't discuss what was going on that very minute in the basement of the Hutch.

There wasn't much to look at; it just looked like water flowing slowly through a tube. But inside the column, antibody-coated B-cells were binding to the avidin-coated beads. The rest of my white-cell fraction—which included my stem cells—continued flowing through the column, past the beads and back into the fluid pathway. In this experiment there were actually three collection bags at the bottom of the device. The unselected cells were collected in the first bag. The washing fluids that cleansed the column after the cells flowed through were collected in a second bag. Normally the entire column with the B-cells still attached would be thrown out, but because this was an experimental process and we needed to determine how effective it had been, the B-cells were knocked off the column by agitation and collected in a third bag. The contents of that bag were analyzed with a FACS instrument to ensure that it contained mostly B-cells, and to get some idea of the number of normal B-cells it contained compared to malignant cells.

Theoretically, when this part of the process was done, most of the biotinated tumor cells had been fished out of my white-cell fraction. But not all of them. As Sharon and Kirsten watched, the blocking reagents were added: first, free avidin; then free biotin. It was a real chemical brouhaha, with everything in the bag bonding to everything else.

Although the blocking reagents had worked in the lab, there was no guarantee they would work in the clinic. Later we heard that other scientists just didn't believe we could get the process to work by doing it this way. I think the term that they used was "baloney." This was pure empirical science. It worked because it worked.

Before introducing the stem-cell antibodies to the solution, Rowley's technicians wanted to get rid of as much of the excess liquid as possible. This was just waste material. So the contents of the bag were put into a large centrifuge and spun through several cycles. Consider it a high-tech washing machine. After several cycles, all the supernatant, the excess liquids, had risen to the top, and the cells, because they were heavier than anything else in the solution, had sunk to the bottom. When this waste material was eliminated, all that was left was the washed white-cell fraction—which contained my stem cells as well as an unknown number of tumor cells.

That material was put back into a sterile bag and biotinated stem-cell antibodies were injected into it. After giving these antibodies about a half hour to find and bind to my stem cells, the bag of solution was hooked onto the top of the Ceprate instrument, the tubing was connected, and, as melting snow flows down a mountain, the solution began flowing through the column.

Sharon and Kirsten supervised the entire process. They helped weigh and measure everything; they did the math that needed to be done. They did everything they were permitted to do. And when Rowley's technicians got hungry, they even bought food for everyone.

The parade of white cells moved slowly through the column. As before, it was impossible to see anything happening. The Ceprate device hummed and occasionally beeped. Once the sound of progress was the roar of an engine; now it's the beep of a computer. Finally a clear liquid began flowing out the bottom of the column into a waste bag. Eventually the humming stopped. The cell selection was complete.

An impeller gently stirred the beads inside the column, causing them to flex. The bond between the antibodies and stem cells is considerably weaker than that between avidin and biotin, so this flexing motion released the stem cells, which were then washed out of the column into another collection bag. The entire process—the preparation, the tumor-cell depletion phase, the addition of the blocking reagents, and the stem-cell selection—had taken about three and a half hours. The experiment had ended.

The results would not be known for several hours. To determine how successful the process had been in eliminating tumor cells, a technician ran a sample through the FACS. The rest of my cells in that bag were concentrated, a very small amount of DMSO was added, and they were frozen in liquid nitrogen.

The fact that all this was going on remained in the back of my mind, but I wasn't nervous. I had complete confidence in these people. I was scheduled to return to the Hutch the next morning to repeat the apheresis process so that Ollie Press would have a reserve in case there was a problem with this collection. But I felt certain we would never need it. From the moment I had been diagnosed with cancer, all the pieces had fit together so perfectly it was as if it were part of a great plan. I had been running the only company in the world working on this advanced treatment for my disease. A team of scientists at that company had successfully completed a year's work in weeks to enable me to be treated. A doctor I'd met briefly in court had developed perhaps the most aggressive treatment for my particular disease. I had absolutely no reason even to consider that at this point something would finally go wrong.

Once again, I'd forgotten to watch for hidden reefs.

A scientist named Mike Loken, an expert in FACS analysis, would do the final analysis. Using his specialized FACS system, he would be searching for the presence of two B-cell markers, known as CD-5 and CD-20, which are highly expressed in mantle cell. There are a lot of them. While it was important that the presence of all the other tumor-cell indicators went down, we wanted to eliminate completely CD-5 and CD-20. We were trying to get to zero.

Tuesday morning I returned to the Hutch and donated a second bag of cells. Later that day, Sharon and Kirsten returned to CellPro with all the data and a small sample. Late in the afternoon they began running a few assays. The initial results looked promising. But we weren't set up to run the extremely sensitive tests for the CD-5 and CD-20 markers. For the final result they would have to wait for Dr. Loken to finish his tests.

In his laboratory, Dr. Loken was disappointed to see little dots popping up on his monitor. Each dot was a tumor cell, boasting that it had survived. Scott Rowley called Joe Tarnowski with the news.

Patty answered the phone when Joe called. I heard only her side of the conversation. "What does that mean?" she asked. "Is that good?" The apprehension in her voice told me the news was not good.

The number of tumor cells had been reduced by more than three logs, more than 99.9 percent. That might seem impressive, but it wasn't good enough. A sliver of difference can make the difference between winning the Kentucky Derby and being an also-ran, and it could make the difference in my fight for life. "It's very good," Joe had explained to Patty. "We decreased it a lot, but it's not necessarily where we want to be. It's still not zero."

I heard her ask, "Can we make it better?"

Finally I got on the phone. The numbers were very good, Joe was right about that, just not good enough. There were more than enough stem cells to go ahead with the transplant, though I'd never doubted that, but Loken's tests had discovered that the system had reselected for tumor cells. So there might also be enough tumor cells for my cancer to return. No one knows how many damaged cells are one too many. Listening to Joe explain exactly what the numbers meant was like hearing early reports of a hurricane heading your way. The storm might veer off or hit head-on, there was absolutely no way of predicting the future, but it made sense to prepare for the worst.

After hearing the full report, I told him, "Listen, buddy, if we've got a shot to make this better, I'd really like to take it."

"You know, Rick," Joe again pointed out, "these numbers really are pretty good. They did a great job." In scientific terms, the experiment had been a great success; the overall B-cell depletion had been extremely good. But in personal terms it had failed, since the CD-5 and CD-20 subsets had not been completely eliminated. This was the experimental version of the cliché: the operation was a success, but the patient was going to die. It had succeeded in the fact that important information had been gained. We'd made more progress toward the development of a B-cell depletion system in weeks than we'd previously been able to make in years. This was clearly the best tumor-purging system available anywhere in the world at that time. But not all of my tumor cells had been eliminated.

In the normal experimental process, the next step would have been to try to find out why all the CD-5 and CD-20 had not been depleted. After analyzing the data, Nicole's team would do this by repeating the experiment several times, changing only one small aspect of it each time.

I didn't have time for that. Although I tried to hide my feelings, I was terribly disappointed. As a manager I've always tried to create a work environment in which failure is an acceptable step en route to success. You don't get people to push the envelope if they're afraid to fail. But I just hadn't expected them to fail this time. After a lifetime spent in the world of experimental science, I should have known better than to anticipate complete success on the first try. Rarely, if ever, do you reach the end at the beginning. But, boy, this time that was exactly what I needed to happen.

There had been occasions in my career when I'd pulled rank. I'd insisted that something be done my way because I believed it was the right way to do it and I could make that demand. But I'd never done that for personal gain. This situation was a bit more complicated. Until this point I had been able to make the legitimate argument that the company was developing a potentially very valuable product, and I was simply fortunate enough to be the first beneficiary. But product development did not require repeating the process the next day without even taking the time to analyze the results.

The way I remember it, I remained completely calm as I rationally discussed options first with Joe, then with Ollie. They remember me being adamant that we try it again. I pulled rank. Maybe I never said the words *I want this done,* but I knew that wasn't necessary. Joe was always very good at catching the wind and going with it. "Look," I said, as lightly as possible, "why don't we just take another run at it? I mean, what the heck, right? I've got stem cells coming out of my ears. I mobilized like nobody's ever mobilized before. I've got no plans for tomorrow. I'll just go back and sit on the apheresis machine for a couple of hours and read a book or something."

I was putting Joe in the middle of a difficult situation. We were friends as well as colleagues. He was caught between wanting to please me and dealing with reality. The reality was that no one knew why the experiment had not been completely successful. So there was no valid scientific rationale to try it again before doing a great deal of preparatory work. But neither one of us had ever been in a situation like this one. In fact, I don't know of anyone else who ever had been. For the first time in both of our lives, as we discussed conducting another experiment, we weren't talking about some nameless patients, guinea pigs, whom we would never meet. This wasn't about statistics, this was about saving my life. My feeling was that I had one shot left at the Death Star, and I damned well intended to take it.

It was a long conversation. Joe read me the highlights of the data. Neither one of us could understand what had gone wrong. For some reason the same procedures that had worked so well in our lab had failed when done at the Hutch. "Could be anything, Rick," Joe explained, "there's just no way of knowing until we have a chance to really look at all this."

Obviously it made no sense simply to repeat the experiment without making at least some changes. Doing it again exactly the same way would be the ultimate hubris. We would be blaming the failure of the experiment on nature rather than on ourselves. It would be as if we were giving nature another chance to get it right. The problem was we had absolutely no knowledge on which to make changes in the experimental process. There was nothing in the data that seemed wrong. As far as we could figure out, everything had gone exactly as planned. So whatever we did would be a guess. A shot in the dark.

I wanted to make substantial changes. I remembered a story I'd read about two major-league baseball umpires who'd made a controversial call that had resulted in a batter who had just been called out batting again. Given a second chance, he'd hit a home run. "I'll tell you one thing," one of these umpires had said to the other as the batter circled the bases, "if we're wrong, we're wrong big!"

If I was going to be wrong, I wanted to be wrong big. "What do you think about running it through two purging columns? That'll give us two good shots at getting all the tumor cells."

Apparently, Nicole's team had actually tried that in our lab. The result had been that a substantial number of stem cells had been lost in the process.

I made several suggestions. Reverse the process again, select for stem cells first, but continue to use the blocking reagents. Slow the flow rate. Joe listened dutifully, making some of his own suggestions. There was nothing more that could be done that night. Joe would speak to Nicole the next morning and we would decide how to proceed. But I had little doubt that we would be trying again.

As it turned out, the timing could hardly have been worse. Weeks earlier, Nicole, Sharon, and Kirsten had registered for a one-day seminar to be given that Wednesday by a noted Yale professor in Bellevue, Washington, on how to present data to enable other people to understand it. This was an important seminar. Data communications is a persistent problem in science; numbers can mean different things to different people, and that can be disastrous in a business so dependent on precision. At times it's like trying to understand someone speaking with a thick accent.

Nicole had taken her pager with her to Bellevue, "just in case," as she'd told Tarnowski, "there's some issue we need to discuss."

To make the situation even more complicated, Scott Rowley's lab at the Hutch was packing up and moving into a new space that day. All their equipment was in boxes, nobody would be able to find anything, so getting any work at all done on Wednesday would be impossible. I would have to wait at least one more day.

It was almost midnight when I finally got off the phone. I was physically and emotionally exhausted. But I lay awake for a long time that night, a long time, trying to figure out what had gone wrong.

On Wednesday morning Nicole was sitting calmly in the auditorium at Bellevue's Meydenbauer Center when she was paged by Joe Tarnowski. When she realized Joe was trying to find her, she was curious, but not the slightest bit alarmed. The call could have concerned any of the projects in which she was involved.

Joe managed to stay on top of just about every project in the shop. During the rush of the last few weeks to complete the lymphoma-purging project, several other projects had fallen slightly behind schedule. As far as Nicole knew, the Rick Project had ended, at least temporarily, in Rowley's lab. The last results of the experiment she had seen had been quite promising. Now it was time to play catch-up with some of the other projects.

During the lunch break she returned Joe Tarnowski's call from a pay phone in the lobby. "I've got some results for you," he told her. "They're not as good as we'd hoped. Rick sounded pretty upset about it."

The lobby was crowded and noisy. Nicole huddled in a corner, trying to hear a little more clearly. As Joe read her the final results of the tumor-depletion phase, she jotted down the numbers on a large sheet of scrap paper.

Ironically, she'd just spent the morning listening to a noted expert explain how the same numbers can have very different meanings to different people. The numbers she'd written down on the back of an envelope were a perfect example of that. To Nicole, those numbers looked pretty good. I'd started with several billion tumor cells. The process had successfully removed three logs, almost all of them. She focused on the 99.9 percent that had been depleted, I focused on the 0.1 percent that remained. "Gee, Joe," she said, "these look pretty good to me. What does Rick want us to do?"

"He thinks you can do better," Joe reported. "He'd like you to take another shot at it."

"Oh, no." Nicole paused to take a deep breath. It wasn't fair, she thought, the team had worked too hard for too long. They were all exhausted. She glanced at the numbers again. The team had done an incredible job. Nothing had blown up, and they had successfully concentrated more than enough stem cells to proceed with the transplant. Results like this would have been acceptable for most experiments. More than acceptable. Besides, as everyone knew, the detection system wasn't perfect. These numbers might even be better than they looked. Once, when she was in college, she'd forgotten about a loaf of bread left baking in the oven. It had been burned rock solid, and she'd almost started a fire. That was a real failure, the kind of disaster she could hold in her hands and no one could dispute. But these results were good. If she had been baking a loaf of bread, people would be complimenting her. "Come on, Joe," she urged, "look at these numbers. I don't know that we *can* do any better than this. Can't you just explain to him that he's got a finite number of tumor cells? Even after he goes through chemo and radiation, he's still going to have some of them in his body. Can't you talk him out of it?"

"He's pretty adamant about it," Joe said softly, almost as if he hadn't heard her. "He thinks maybe we can do two purging columns."

Nicole continued staring at the figures she'd written down. Ninety-nine-point-nine percent. "No, we've tried that," she responded somewhat distractedly. "It doesn't help. The second column just ends up depleting more stem cells." The numbers began taking on meaning. She began seeing ratios in her mind.

Joe continued, "What if we switch the process and do the stem-cell selection first? That'll decrease the population of—"

"No, no way, Joe, we're not going back to that," Nicole said, anger trembling in her voice. After all the work the team had done, she wasn't going to allow anyone to tell them what to do. Just what did Rick expect, she wondered, perfection the first time out? But as she wrapped her mind around those numbers, her curiosity clicked on as if someone had flipped a switch. There were a lot of variables in the process, but some of them impacted the results far more dramatically than others.

Tarnowski would not demand that Nicole run the experiment again. That wasn't his style. Instead he suggested it about as forcefully as I had suggested it to him. "Rick's back at the Hutch right now. He's getting another apheresis done this afternoon."

Nicole sighed. "I just don't know, Joe," she finally responded. "I'll talk to everybody, but you tell Rick my off-the-cuff assessment is that he's done pretty well. He got rid of three logs. We did pretty good." But even as she said that, the pull of the numbers grew stronger.

Sharon and Kirsten had been waiting patiently for her. "You guys aren't gonna believe this," she said, "but Rick wants us to do it again. He didn't think the numbers were good enough."

The three of them huddled over the numbers Nicole had scribbled onto the back of the envelope. "Ah, shoot," Sharon said, "this looks like it's reselecting."

"So what do you want to do?" Nicole asked. What she meant was, *Are you willing to try it again?* If they agreed, the task facing them would be herculean. First they would have to try to figure out what had gone wrong and then determine what to do to fix it. Then they would physically have to make all the preparations to do it. They'd have to mix new blocking reagents. Scott Rowley would have to get his staff together. Rowley's lab had been packed up. They would have to find the equipment they needed, pull it out of boxes and set it up, and make all necessary adjustments. Even given several days, getting it done successfully would be on the distant side of the possible. But they didn't have days. The experiment would have to be planned, prepared, and done within a few hours.

As they stood in a small group outside the auditorium, I was in the middle of a third white-cell apheresis. This was my last chance. The mobilization window remains open only briefly; three days was a stretch. My bone marrow had pumped out about as many stem cells as it was capable of producing. I had three bags full.

Nicole did not want to repeat the experiment. For months the team had been working under the pressure of a ticking clock. If they went ahead with this, their time would be measured with a stopwatch. Literally seconds would count. Outside the body, cells start deteriorating almost immediately. After eight hours they tend to get very sticky, making it difficult to work with them. Within a day, they die.

The Rick Project team didn't have time to waste discussing whether or not they had enough time to complete the project. They just leaped in and went right to work. Among the three of them, they had conducted literally thousands of experiments in their careers. They'd probably spent half their adult lifetimes in laboratories. Each of them had enjoyed numerous successes and suffered countless failures. They understood the chemical bonds that held the world together, and the cells that were the stuff of life. They spoke that esoteric language of numbers. And everything they'd done in their careers, every test tube they had filled, every solution they had measured, every chemical reaction they had done, all the years spent memorizing formulas and staring into microscopes, the days and weeks and months spent sitting at a lab bench, all of it came together in the lobby outside the Meydenbauer Center.

It would be just another experiment done in a lifetime of experiments. But it would be the one experiment none of them would ever forget.

It would not be much of an experiment in the traditional sense. It wasn't part of any natural progression. This was hold-your-nose-and-jump-in-the-water science. It was clear that B-cells were sticking to the stem-cell selection column. As the three women looked at the numbers, it became clear that there were a least a dozen different reasons this might be happening. The problem might be solved relatively simply by adjusting any of them, or perhaps—and this would make it far more difficult—it might require changing two or three or more different elements. With this experiment we were attempting to manipulate nature to our human needs, and there was no way of knowing what lay hidden before us.

Nature sometimes makes itself known with triumphs like a rose in bloom, but more often it keeps its secrets as silently as the movement of an atom. No one on the team really knew how all the molecules they were throwing into the pot reacted to each other. Nature had rules yet undiscovered. It was possible the experiment had worked as well as those rules would allow. Early in CellPro founder Dr. Rich Miller's career, he had seen antibodies destroy tumor cells in a patient. Wipe them out. For a brief time he'd believed he had found the path to the cure for cancer. But that path led nowhere. The results could not be sustained. When the tumors returned, the antibodies were no longer capable of destroying them. Nobody understood why. That's just nature, they said. Just nature.

As Nicole, Sharon, and Kirsten reviewed their many options, they started making notes on the back of the seminar program. Mark Lodge, an associate scientist at CellPro also attending the seminar, joined the discussion. They could do as I had suggested and run my cells twice over the same depletion column, or run it through two different columns. They could change the concentrations of the blocking reagents or add more antibodies. They could even eliminate completely the B-cell purging step and try multiple runs through the stem-cell selection column. The problem could be the flow rate through the column: too fast, and the cells would not have enough time to bond; too slow, and B-cells would bind to the stem-cell selection column.

Hundreds of people were noisily milling around them as they worked. But the team was almost oblivious to their presence. Their concentration on the problem was absolute. A decision had to be made, and it had to be made at that moment. Each of them knew that the answer was hidden in the numbers, it had to be there, but it was as well camouflaged as a chameleon at night. "It has to be the blocking reagents," Nicole said, "it just has to be. Either one of the reagents wasn't any good or we just didn't have enough of it."

They knew the reagents worked to some degree, Kirsten pointed out, because they had gotten some B-cell depletion. Just not enough.

"Well, how much more do you think we need?" Sharon wondered.

They debated the numbers. How much avidin and biotin were needed to sop up all the biotinated B-cell antibodies? A delicate balance had to be found. Not enough, and the tumor cells would be reselected. Too much, and the excess would poison the column, preventing stem cells from binding to it. There is an old cliché about three women who have had lunch together dividing the bill to the penny and asking, "Who had two iced teas?" This was what had become of that tired joke, three very bright scientists debating the number of moles, the measure of molecules, that were needed to save an experiment and perhaps a life.

The discussion continued for twenty minutes, then a half hour, forty minutes. Time stood still even as it raced away from them. Because there was nothing in the numbers to indicate that the ratio between the avidin and biotin had caused any problems, they finally agreed to maintain that. In the next experiment, they decided, they would simply increase the amount of the blocking reagent. They would add a lot more avidin and biotin.

Science is often inexact, but rarely are vitally important decisions made on such flimsy evidence. If the numbers were accurate, they could increase the amount of blocking reagents tenfold without interfering with the stem-cell selection column. Doubling the amount would be safest; ten times was reaching the upper margins, but, according to the numbers, would still be safe.

They finally reached an agreement that they'd compromise. They'd repeat the experiment, but bump up fivefold the amount of blocking reagent to be added. That number had all the scientific validity of a lottery ticket picked out of a hat. Later, Nicole explained that the team had analyzed all the data and then taken what I refer to as a SWAG, a "scientific wild-ass guess." But that was the decision: increase the amount of blocking reagent fivefold. And then they would cross their fingers and just hope that it worked. They stood there in a little group, repeating that number over and over to ensure that Sharon and Kirsten would remember it when they returned to the lab that night. On the back of a seminar program the numbers made complete sense, but all of them knew that cancer had been cured many times on a sheet of paper.

Although it was my life, there was no reason that I needed to be told about their decision—beyond the fact that they would conduct another experiment Thursday morning. "They think it might be the reagents," Joe had told me. "They're going to increase them a lot."

"Yeah," I said, "that makes sense." While I was not a scientist, I understood the science. It did make sense to me. But so did changing several other variables.

Instead of driving home from Bellevue that night, as they had planned, Sharon and Kirsten went directly to their lab at CellPro to gather the equipment they would need the next morning to brew up another magic mix of free avidin and free biotin. Whatever happened the next day, I would have to live with it, or perhaps die because of it.

Experimental science isn't generally so dramatic. Advances are slow and progressive, rather than sudden and drastic. The "eureka" moments of scientific history, like Alexander Graham Bell ordering his assistant, "Mr. Watson, come here, I want you," and discovering his telephone worked, just don't take place anymore. The primary reason biotech companies now have such difficulty attracting investment capital is that it takes so long to move from an idea to an approved product. Usually it isn't enough just to get good results; you have to understand the reason for those results, and document them, and repeat them. But on this Wednesday night in June of 1996, to the statement that good scientific research consists of equal parts inspiration, dedication, and perspiration, it would have been correct to add one more word, *desperation.*

Early Thursday morning, Sharon returned by herself to the Hutch with extra columns, vials of antibodies, and newly mixed blocking reagents. In her pocket she also carried a pager. She hadn't wanted to take it—she was afraid Joe would spend the morning paging her to find out how the experiment was progressing—but Nicole wanted to be able to reach her in an emergency.

Rowley's lab was in chaos. Boxes were stacked in piles all over the floor. Before they could even begin to process my blood, they had to spend several hours setting up the equipment.

Sharon had been right; several times during the morning, Joe had paged her, but she hadn't returned his calls. As the morning faded into the afternoon, Joe was getting very worried. Silence was usually a sign that there were some problems.

Nicole sat in her office at CellPro, working on the breast-cancer purging project, actively trying not to think about what was going on at the Hutch. In the lab, Kirsten glanced occasionally at the clock, wondering how the experiment was proceeding. I waited at home, pretending to read. I found myself rereading the same material several times without comprehension. I decided I was really tired. I knew we would not be getting any results for several hours, but I was disappointed that Joe hadn't called to give me a report.

My apheresis collection, the cellular component of my blood, had been out of my body overnight. A long time. My cells had gotten very sticky, making it more difficult for them to flow easily through the system.

Joe's frustration grew as he continued paging Sharon without receiving a return phone call. There was simply no way for him to know that in the subbasement of the Hutch the pager did not work. Sharon never received his page. So she worked uninterrupted through the morning.

The experiment proceeded as planned. My sticky cells caused more problems than anyone had anticipated. They had formed thick clumps, making it very difficult to run them through the system. Sharon suggested they split the bag of cells in half and run the collection through two different purging columns. A second Ceprate device had to be quickly set up.

Once the columns were loaded, the device practically ran itself. About the only thing Sharon had to do was make sure the technicians used the blocking reagents in the correct order. If they added the avidin before the biotin, the experiment would be destroyed. It seemed like a difficult mistake to make, but unfortunately that is precisely the kind of mistake that gets made in labs. Rowley's people were good technologists, and they followed the recipe to perfection.

The processing was completed by midafternoon. Loken's FACS had become my crystal ball. Whatever it saw would foretell my future. Either I would receive a transplant of pure uncontaminated stem cells, in which case I would probably get to watch my boys grow up, or tumor cells would be put back in my body, and then my prognosis might not be so promising.

Loken sat in front of the FACS, watching my cells flowing through. And after a long wait, when he was sure of his results, he smiled.*

I waited. On Friday morning, Scott Rowley called Ollie Press with the data. "This looks pretty good," Rowley said professionally. "We've got enough stem cells." In fact, the results had been spectacular. The most sensitive FACS scans detected no tumor cells. None. By increasing the blocking reagents fivefold, they had depleted an incredible 4.5 logs of tumor cells. That seemed almost impossible; even intricately planned experiments rarely worked quite so well. But the numbers tell a story without emotion. No detectable tumor cells.

The calls started coming a little faster. Scott Rowley called Nicole and read her the data. Nicole leaned back in her chair and shook her head in wonderment. It had worked, the whole shebang had worked. Then she walked rapidly down the hall to share the news with Sharon and Kirsten. "Listen to this, you guys aren't gonna believe this one," she said. "We guessed right. They can't find a tumor cell." Ironically, when she called Joe Tarnowski he wasn't in his office, so she left him a voice mail.

Ollie Press called me as soon as he hung up with Rowley. Ollie is not given easily to emotion, but this time he couldn't help it. "Wow, Rick," he said, "I just heard from Scott. We got a lot better results on this run. It looks really good."

"Boy, that's great," I said. "I didn't know that. I haven't heard from anybody."

---

*Ironically, Mike Loken became an expert witness for Baxter against CellPro at our second trial in Delaware.

Ollie continued, "Yeah, it looks like we're ready to go here." The happiness Ollie felt wasn't just for me. If these results could be repeated, the world of transplants had just taken a significant step forward. The B-cell depletion system could help save the lives of countless lymphoma and leukemia patients. It would help those kids I'd seen in the apheresis unit. It meant that autologous transplants could be done without fear of reinfusing tumor cells.

Joe Tarnowski called me later that morning with more details. "Gee, Rick," he said, "I gotta tell you, we really looked and we couldn't find anything here. It looks like we got them all."

Nicole found Stan in Dr. Elizabeth Sphall's lab in Colorado. Technicians in that lab were startled to hear Stan suddenly scream, "Yes!" and thrust his arm triumphantly into the air.

I hung up the phone and told Patty. "That sounds better," she said calmly. I let out a huge sigh of relief. Now we had the best possible product put away in preparation for the transplant. Nicole and the Rick Project team had accomplished the near impossible, but their job was done. For me, the hard part was about to begin.

Nicole Provost is not a jump-up-in-the-air-and-cheer kind of person. She took a few minutes to enjoy the satisfaction of a job extremely well done, and then immediately began worrying again. This wasn't really much of an experiment, she knew, and she was afraid that I would immediately want to turn it into a product and get it out to the marketplace. That wouldn't work, she planned to tell me. The column needed to be changed drastically, there were too many variables in the recipe, success depended on technicians doing everything exactly right. She was sure the FDA would never approve the depletion column as it existed. An incredible amount of work would have to be done before it was even ready to begin clinical trials. It was bad product development, she knew.

But she also knew it was good business. And so she began making detailed notes of what changes had to be made in the system.

I began preparing both mentally and physically for the actual transplant as I had prepared for few other events in my life. I was determined that nothing, absolutely nothing, was going to get in my way. Larry Culver was doing a fine job of running the company, I didn't need to worry about that. I was getting back pure stem cells, I didn't need to worry about that. I was focused on the immediate future.

So I wasn't prepared for the blow that hit me. In my corporate life,

I've made and received thousands of telephone calls. In total, I've probably spent years on the phone. Very few calls stand out in my memory. This one did. It was a call I will never forget.

It was mid-July and Coe Bloomberg, our lead attorney in the lawsuit, telephoned from California. "It's nice to speak to you, Rick," he said, "How are you feeling?"

"I'm doing okay," I responded. "I've got one more big treatment to finish, but I'm hanging in there."

I heard him sigh. "Well," he said, "I'm afraid I'm going to put you through one more big challenge. The judge just overturned the jury verdict."

"What?" I said. I knew that couldn't be true. Throw out a jury verdict? That wasn't possible. Not in America. Trial by jury was right there in the Bill of Rights. We'd had a fair trial and won. It was over, done, finished. I was fully aware that there would be appeals. I knew it was possible that some elements of the appeal might be granted. But no one could possibly have anticipated a far-reaching ruling like this.

In a fifty-eight-page opinion, McKelvie had decided to reinterpret Civin's original patent claims, which he was entitled to do under the *Markman* decision, and on the basis of his new interpretation he ruled that CellPro had infringed two patents and ordered a new trial on the other two. Basically, he decided that the facts of the case had been too complicated for a jury to understand, causing jurors to reach a decision that was not supported by the evidence. It was a stunning decision. I had believed I was long past that point at which I could be surprised by anything that happened in my life, but this . . . this was incredible. After hanging up with Coe Bloomberg, I sat at my desk, staring into space. Oddly enough, I wasn't really angry. I guess I was incredulous. Everything we had gone through with Baxter up until that moment, the years of negotiations, the bitter lawsuit, the trial, were suddenly ancient history. Bloomberg's firm, Lyon & Lyon, had tried thousands of patent cases, and they could not find a similar decision anywhere in their records. McKelvie had put the ball back in play, and as it got closer it began to look like a wrecking ball.

This decision convinced Josh Green that Judge McKelvie really had it in for us, and that we had no chance of winning. "It's over," he said. "He's going to make sure we lose. And we're going to hell shackled."

Lyon & Lyon immediately filed an appeal. Hearings were scheduled for the fall. Although I couldn't understand how a judge could wield so much power after a trial had been concluded, there was nothing I could do about it.

I was devastated, absolutely devastated. It was worse than our worst fears, because it was something we had never even contemplated. The fact that it had happened was unbelievable, but the timing made it even more terrible. I recognized immediately that resolving this was going to require the same level of activity we had brought to it for the first trial, maybe more, and at that moment I couldn't do it. I had no options, no choices. I felt as if I were abandoning the company at a time when I was really needed, but as much as I wanted to get up and lead the charge, my job was to get this transplant done.

Truthfully, I also felt a little bit relieved that I didn't have to go back up on the front lines for a while. I really needed the rest.

I had been warned by Sherri Bush that getting back my cells would be far more difficult and painful than donating them had been. Essentially, Ollie Press would be detonating the equivalent of a microscopic atom bomb in my body. I would receive enough radiation and chemotherapy to kill every rapidly dividing cell in my body, from my hair follicles to the lining of my digestive system. It was as if a house were being emptied of its contents, including the paint on the walls, from the inside.

The first part of the fight for my life had been more successful than anyone but I had dared imagine. When the time came for my transplant, I would be receiving an infusion of tumor-free, purified stem cells. But the work done by the Rick Project team would be valuable only if Ollie Press was able to kill all the tumor cells growing in my body.

The two proven methods of killing rapidly dividing cells are radiation and chemotherapy—two different roads to the same destination. Historically, most transplanters have treated patients with a low dose of radiation, delivered to the entire body, combined with chemo. Ollie Press's high-dose radiation protocol was completely experimental. And extremely controversial. Although the program was sponsored by the National Institutes of Health, it had not received FDA approval.

Ollie wasn't interested in debating the merits of his program. Too often as a young oncologist he had seen cancer patients receive enough external radiation to put their disease into remission, but not enough to destroy all the mutant cells that caused it. Their systems just couldn't tolerate any more. Eventually their cancer had returned. And because only the toughest cells had survived the radiation and chemotherapy, the disease returned more resistant to treatment and virulent than it previously had been.

It was obvious that the more radiation a patient tolerated, the better his or her chance for survival. The problem with total-body irradiation, in which the patient's entire body is exposed to radiation, is that often the lungs, liver, and the kidneys take pretty direct hits and, over time, begin to degrade. In many cases the patient eventually suffers long-term disability or dies from the cure. So, rather than trying to find a way of increasing the amount of radiation that could be delivered by total-body irradiation, Ollie began using monoclonal antibodies to deliver massive doses of radiation directly to tumor cells.

By the time I entered his program, he was routinely giving patients the highest doses of radiation available anywhere in the world. Radiation is measured in curies. In most treatment centers, for example, thyroid cancer patients receive up to 100 millicuries. At the University of Washington, thyroid cancer patients commonly were receiving as much as 800 millicuries, a staggering amount.

Nature's delivery system is exquisite. A tumor cell has between 15,000 and 200,000 potential antibody binding sites, so, in theory, numerous different antibodies could be used effectively to target it. The antibody Ollie used in the high-dose protocol, called CD-20, binds to a molecule found on more than 95 percent of all lymphoma cells. That means that the antibody doesn't have to be custom-tailored for each treatment; it will bind to the tumor cells of virtually every lymphoma patient.

The maximum amount of radiation safely delivered by external beam, the usual type of radiation given to cancer patients, is no more than 1,500 centirads. After running all the tests, after measuring the size of my lungs and kidneys and liver and spleen, the nuclear medicine technician calculated that I should receive 581 millicuries, which is about 2,300 centirads. That much radiation delivered by external beam would have killed me. That's a fatal dose, more than enough to blow the needle off a Geiger counter. I wanted as much radiation as my body would tolerate. I wanted Ollie to clean house.

I was going to be radioactive, much too hot to be around other people. So until I cooled down, which would take at least ten days, I would have be isolated. I would have to stay in one of the University of Washington's two specially built rooms. The cost of the radio-labeled anti-body treatment is about $20,000. Patients without the proper insurance— people from other countries, for example—had to put the money on deposit before entering the hospital. Survival is expensive.

I checked into UW Medical Center on August 2. Jamie had flown in early from his summer course in Venice, Italy, that day to be with us, so I went to the hospital with Patty and my boys. It was almost as if I were checking into a fashionable hotel rather than beginning the most grueling month of my life. I carried with me into isolation my computer, my Disc-man and a selection of classical music, as well as a substantial amount of paperwork and several magazines and books, among them a book titled *Adrift*, the story of a man who had survived in a raft on the ocean without food or water for twenty-five days.

CellPro had hired a video cameraman to record this scene. Maybe I was a CEO, but at heart I was still a marketing guy. A salesman. And I wasn't too sick to recognize a great publicity opportunity. The fact that CellPro's Ceprate system and CellPro's brand-new B-cell depletion device— which would be available as soon as possible—had been used to help save the life of the company's CEO would make a great marketing tool for those products. Assuming they did save my life, of course.

We'd been documenting the entire procedure for several weeks. The day before entering the hospital, I'd taped a long interview. "I'm going to be involved in experimental radiation treatment," I'd said quite confi-dently as I looked straight into the camera. "I'm essentially radioactive for ten days or more. Because of that, they have to isolate me in a lead-shielded room. Nurses won't even be allowed in. I'm in complete isolation for ten days. I'm taking in everything I'm going to need for ten days. As you might imagine, I'm taking a lot of books. I've got a stack of books this high that I have been dying to read, and I never seem to have time to read anything but technical material, so I'm really looking forward to doing some of that. And I'm taking my computer. There'll be a phone in there, so I'll be spending a lot of time on the telephone, which I'm used to. . . . I'm told you don't feel particularly ill during this period of treatment."

While some of that was certainly bravado, most of it was true. I was very optimistic as I began the transplant procedure.

Sherri Bush had prepared room 6318 for my stay. From the outside it looked just like an ordinary hospital room; what made it unique were the six-inch-thick lead shields that had been built into the walls. As she had explained, I would be secreting radioactivity through the sweat glands in my hands and feet, so everything I might touch while I was in the room had to be protected. Every single thing that came into the room while I was there, the food I didn't eat, the plates, the glasses—everything— would have to stay there. Patty and I watched as Sherri covered the floor and much of the walls with barrier paper that would capture radiation. The top layer is about as porous as a diaper; the bottom layer is sealed to prevent moisture from seeping through. Then she wrapped in protective plastic the telephone, my computer, everything I intended to take out of that room with me. When I finally walked out of that room, anything that had not been protected would have to be sealed in a lead-lined container until it was no longer radioactive.

In the six months since I'd noticed a small lump in my neck, I'd come all the way from "take two aspirin and call me in the morning" to nuclear medicine technologists, radio-labeled antibodies, and Geiger counters.

Patty and I wanted to spend a few minutes alone before the proce- dure began. I hugged each of the boys, and they went off to do some work on the boat. Patty and I got about two minutes together. What do you say in two minutes to the woman with whom you've spent about half your life? We hugged each other. "Okay," Patty said confidently, "I'll see you on the other side."

The radioactive material was a clear liquid. While it looked harmless, if you drank it your hair would fall out and then you would die. This material was produced in a cyclotron in Canada, and UW had purchased it from New England Nuclear in Boston. It was wheeled into my room encased in a lead-lined container known as Big Bertha. It was actually sort of funny to see little Sherri wheeling this large machine into the room. "Okay, here we go," she said, as she connected the bag of radioactive solu- tion inside Big Bertha to my Hickman catheter and began pumping this liquid into my body.

"In case you forget," I told her, "I'll be the one wearing the glow-in- the-dark underwear." We both laughed.

I didn't feel anything at all as she loosed these antibodies on their search-and-destroy mission. The process took about an hour. When it was done, the nuclear medicine technologist, Larry Durack, stood several feet away from me and turned on a Geiger counter. It started singing. "Man, you are really hot," he said. "I don't remember too many people ever being this hot."

I didn't know if that was good news or bad news. I just lay in my bed and closed my eyes and waited for whatever was going to happen. But inside my body an extraordinary battle had begun. This was an invisible war, beyond even the imagination of George Lucas. Billions and billions of antibodies carrying ray guns of radiation were sailing silently throughout my body, searching for tumor cells. When an antibody found one of those cells, it locked onto it firmly. And then, like an alien presence, the radiation destroyed the machinery of the cell. It broke up the DNA, preventing the cells from functioning or replicating. And eventually, like a dead planet sailing aimlessly through space, that cell continued through my system until it was removed by the spleen.

I knew this had begun to happen, but for a brief time I felt nothing unusual. Then the first wave of nausea swept through me.

The nausea never went away. If I moved around at all, I felt terrible. There were several tasks I had to do every day, like checking my vital signs or plugging an IV into the catheter. I could do those things, but when I finished them it felt so good to lie back down and close my eyes. All my grandiose plans disappeared. After a day I stopped listening to my music. I never touched my computer. I rarely used the telephone. Using the bathroom—which I could do—required a major effort. At times I thought about CellPro, I thought about the people there and the lawsuit, but there was nothing I could do about any of it. I had absolutely no appetite. After the fourth day I stopped eating. I flushed the food down the toilet and piled up the dirty plastic trays in a corner of the room. Just thinking about food nauseated me. One day I tried to eat an apple, but I couldn't even keep that down. The book I'd brought into the room with me, the story of a sailor adrift on the ocean without food or water, turned out to be the worst possible book I could have chosen. It was sort of like reading about a plane crash while in flight. It reminded me how terrible I was feeling, then magnified those feelings, so I laid it down and never picked it up again. All I could do was lie in bed with my eyes closed.

Emotionally I probably felt even worse than I did physically. This was almost certainly the longest period I'd ever gone in my life without feeling the touch of another human being. I missed that perhaps most of all. I felt alone and abandoned, as if life were passing me by. At times I got totally despondent. I would look out my window and see the sun shining brightly and be reminded that this was summer in Seattle, the best of all times there, and I wished desperately to be out on Puget Sound, sailing north with Patty and the boys toward Murdock Point.

I tried to fight those feelings by focusing on the future. "When I get out of here," I promised myself, "we're going to spend a long time in the Caribbean and I'm going to lie on a boat and relax and do all the things I love to do." I pictured all of us on that boat, luxuriating in the sun, sailing through calm seas.

The days eased one into the next. Once every day Larry Durack would stand in the doorway and take my reading with his Geiger counter. There was a string attached to the end of the counter, and I would have to pick up the string and hold it taut against my chest, to make sure we were measuring from the same distance. Early in the process, Larry had predicted I would have to remain in isolation for twelve days. Sherri told me he was never wrong. So I began counting off the days.

Patty and the boys would visit every day. Sherri discouraged them from staying too long because of the potential exposure to radiation. The door to my room, which was probably twenty feet away from my bed, was left open, but was blocked by a lead shield. A second shield was set up in the hallway, and my visitors would have to stay behind that one. These shields were on wheels, about four feet high and three feet wide. From my bed I could see above the shield in the doorway out into the corridor. As the days passed and I cooled down, the second shield was moved closer and closer to the door. The boys would stay only a few minutes, but Patty would sit down behind the second shield and stay there until asked to leave. Our conversations were about as normal as a husband and wife might have with about a foot of lead between them. How are you feeling today? I paid this bill. My mother called. One thing she never mentioned was how difficult all of this had been on her.

I learned that only later. When we were together, I never saw even the slightest crack in her demeanor. Nothing shook her up. As she had promised at the very beginning of this journey, she maintained. But that was for me. When she was at home, apparently things were quite different. It must have been frightening for Patty and the boys. In all the years we'd been together, she had never seen me so vulnerable. And my kids . . . I know it was difficult for them seeing their father in such a dependent state. And though they never mentioned it, the possibility that I might die must have touched their minds.

Patty, Jamie, and Ben relied on each other for emotional support. The only time they felt free to express their real feelings was with each other. They always put on a great front for me, but in fact they were hiding a tremendous amount of frustration. Adding to the tension was the fact that Ben was preparing to leave for his freshman year at Bates College. At times they argued—"raw moments," Ben called them. Their fights were almost always about something inconsequential, but they enabled them to safely release the tension they were feeling. Patty found the peace she needed working in her garden, and Ben and Jamie found it working on our sailboat. Patty wanted them to stay close to home, but while they helped her with the daily chores, they also wanted to lead their own lives. "You're just hiding your feelings," Patty insisted. "You're trying to go on like nothing's wrong."

Ben's response was the ancient cry of the teenager: "You don't understand."

But one day Ben was driving out of our development, on his way to spend the weekend on the boat with a friend, when he passed his mother returning from the hospital. When they stopped to talk, Patty broke down in tears. After a few moments she went home, and Ben started to go to the boat. But then he turned around and drove home. They held each other for a long time.

Jamie and Ben actively continued to ignore reality. Everything in their lives to this point had gone very well, and they believed unwaveringly that I was going to be fine. Jamie later confessed that sometimes when he was alone, he couldn't help but consider the grim possibilities. "I had this conversation with myself," he told me, "in which I set up this perverse dare. I dared the world, go ahead, take away my hero, and watch how badly it'll screw me up."

Patty also had the company of our cat—I refer to her as "our" cat, but she's always been partial to Patty. Her name is Meghan Marie, but I've always called her Fat Ass. Maybe that's why she's partial to Patty. We'd gotten her as a kitten in 1985, and she'd traveled the world with us. She often sailed with us, and twice, while chasing moonbeams on the water, she'd gone overboard. Patty had heard her cries and saved her. Fat Ass has always been very independent. She's always done what she wanted to do. And while I was in the hospital, what she wanted to do was be with Patty. She obviously sensed that Patty needed her, and at night she would sleep on the empty side of the bed.

The days remain a blur in my memory. The only time I felt comfortable was when I was lying motionless in bed. I suppose some people might have used this time for reflection—to consider the path they had followed and wonder where it might be leading. Maybe even to think about the meaning of their lives. Other people might have used the solitude to make plans for the future. But my mind just did not take me to those places.

I maintained.

Nuclear technologist Larry Durack had hung a chart in my room so I could keep track of my radiation level. Just as he had predicted, after twelve days I was no longer dangerous to people around me. Ollie allowed me to go home for one night before beginning chemo. I was incredibly weak, but I had little difficulty walking out of that hospital. Jamie and Ben drove me home. Although I hadn't eaten in almost two weeks, I still had no appetite. The boys picked up a pizza for dinner, and I took one look at it—and an amazing thing happened. When I was in that room I couldn't eat a thing, everything made me sick. Yet when I got home that night, in familiar surroundings with my family, I could eat without nausea. I wolfed down two pieces and felt wonderful. That, I recognized, was a huge lesson about the power of the mind.

After dinner we went upstairs and watched a favorite family movie, *Back to the Future*. We hadn't seen it in several years, but once we got into it, we started shouting the lines along with the characters. The movie was funny, but the night was spectacular. I had been a prisoner for twelve days, and suddenly I was back with my family, doing the things we loved to do together.

Sherri Bush always urges Ollie's patients to involve their families in the process as much as possible, because during the worst times it's vitally important for them to remember why they're going through this hell. If I had needed such a reminder, that night I got it. We laughed until very late, and then I had the luxury of sleeping in my own bed. I think what struck me most was how much I had missed doing all the little things, like making sarcastic remarks to our cat, but then I remembered that it is all those little things, the daily routine, that make up a life.

I returned to the hospital the following afternoon to begin chemotherapy. Elaborate studies had been conducted to determine which drugs would work best against different types of cancer. *Studies.* That word seems so cold now. It's a word I had read or used in so many forms throughout my career. But what it really meant was human beings fighting for their lives. This group of human beings received this drug, that group received another drug. This many patients lived or showed improvement, that many patients died. End of study.

I have no concept of how many people had died determining that the best course of treatment for me was a combination of two drugs, VP-16 and Cytoxan. Mantle-cell lymphoma was such a new classification that I doubt it was too many. I knew Ollie would be extremely curious to see exactly how I responded to these drugs. So I too had become a study.

These drugs are normally given at night because the side effects are so unpleasant that the hope is the patient will sleep through some of them. I was given the VP-16 the first night. It has to be administered very slowly, so it took almost four hours. I slept through much of it. I was given drugs to calm me down and fight the nausea, but the next day was definitely tough.

The second night I was given the Cytoxan, a drug so toxic that while it was dripped slowly through my catheter the doctors inserted a urinary catheter and continuously irrigated my bladder, otherwise the Cytoxan might have burned it out. That's precisely the kind of information learned in previous "studies." Learned, unfortunately, by experience.

This was the final battle between the best weapons of medical science and nature gone awry. The chemo killed all rapidly dividing cells, including the epithelial cells that made up the mucosal lining of my mouth and digestive system. The only blood cells left alive in my body were circulating red blood cells responsible for bringing oxygen to my tissues and taking away carbon dioxide, and barely enough platelets to stop bleeding. But all the linings were gone. Within twelve hours my mouth turned into a huge open sore. It was unbelievably painful. Sherri put me on a continuous morphine drip and gave me a Xylocaine mouthwash that helped moderate the pain.

After I had received my combination chemo, I was told, basically, okay, that's it. It's all in. Now you just have to go through the consequences. They might just have well have said, "Welcome to hell."

10

I WAS GIVEN BACK A FUTURE TWO DAYS after I received the Cytoxan. Assuming the radio-labeled antibodies and the chemo had done their job, my immune system had been completely destroyed. My body was a shell containing my organs and a small river of red blood cells. My white blood cell count had dropped to zero. If this transplant was successful, my purified stem cells would take root inside the marrow of my bones and begin dividing, sending progenitor cells into my bloodstream and body, where they would mature into a healthy immune system.

My second birthday took place on August 12, 1996. There were no guarantees this transplant would work, but without it I would be dead in less than two years. In the past, I would have been confined to a specially prepared sterile room and kept away from people, but studies had shown that in most cases that had not been necessary. The human gastrointestinal tract is far dirtier than any hospital room, and it's extremely difficult to clean it out completely. If the patient's own body isn't sterile, it makes little difference whether the room is immaculate. So instead of worrying about maintaining a germ-free environment, transplant patients are given large doses of antibiotics to combat infection.

Every patient treats this day differently. Families are permitted in the room during the infusion. Although they're not permitted to kiss the patient, Sherri encourages families to hold hands during the reinfusion process. Many families have a little party, complete with a birthday cake, cider, and balloons. Some people select meaningful music to be played while their stem cells are flowing into their bodies. Sherri remembers one patient who listened to a tape of his daughter singing "You Light Up My Life." Almost everybody takes many pictures, pictures of the stem cells, the doctors, and the room, as well as the patient. It is a celebration of new life.

The actual infusion process is a bit anticlimactic. People expect cutting-edge medical science to look more dramatic, as it does on television. The background music should reach some sort of crescendo. Instead, it's rather straightforward. There really isn't very much to see. Patty and the boys—and the video cameraman—were in the room with me when I got my cells back. My cells had been kept frozen in liquid nitrogen for almost two months. They were in five small vials. At one point, Patty asked the nurse, "Are those test tubes made of glass or plastic, in case one of them falls on the floor?"

Those vials were handled as if they were gold bars. It was extremely unlikely that anyone would drop one of them. "They're plastic," the nurse answered politely.

My cells were thawed gradually in warm water until they reached body temperature. Incredibly, as the nurse took them out of this bath, she dropped one of them on the floor. Nobody said a word. It was eerie. The nurse just picked it up and continued with the process. The five vials of cells were injected into a sterile bag and reconstituted in less than half a teaspoon of fluids. If the stem cells in that bag had been concentrated, they would have been barely visible to the naked eye. The entire collection of stem cells would have looked like a tiny white pinhead at the bottom of a test tube. Maybe they didn't look like much, but without them I would live at most twenty days before infection overwhelmed my system. Groggy as I was from the morphine, I still realized that my life was in that sterile bag.

At just about the same time I received my cells, another patient, a woman in the room next to mine, also got her cells back. This woman's blood had not been purified or depleted. Her cells had been processed the normal way and frozen in DMSO. The contrast between the two of us could not have been more extreme. In essence, we were a comparative study.

For something that didn't look very impressive, that white pinhead had changed a lot of lives. To make it possible, CellPro had been created and had raised and spent more than $100 million on research and development. In the seven years CellPro had been in business, investors had earned substantial profits, lawyers had waged battles, scientists had unraveled natural mysteries, and the result of it all was that infinitesimal suspension of cells in the bottom of a test tube. To me it meant life, but it was also the reason that CellPro existed. At that moment, in that small room at the University of Washington, medical science was taking another small step forward.

My cells were infused into my body in an IV solution through my Hickman catheter. The entire process took about ten minutes. I experienced no discomfort at all, no problems.

The cameraman recorded the entire process. I remembered thinking, "Gee, I don't have my hat on, I don't even have a wig on. I must look terrible here." So much for being on the cutting edge of medical science.

The woman in the next room was not so fortunate. She was receiving a big bag of her peripheral blood cells. To protect those cells, a substantial quantity of DMSO had been added. For each liter of cells approximately 100 milliliters—about half a cup—of DMSO was needed. Basically an industrial solvent, DMSO is pretty toxic stuff, particularly for people who are already sick. It often makes patients violently ill. They start vomiting, they suffer cardiac arrhythmia. It has even been known to cause renal failure. While not deadly, it makes an already weakened patient feel terrible for quite some time. It also has a noxious smell that seems to emanate from your pores and just permeates a room. This woman's infusion did not go well at all. After receiving her first infusion of cells she became extremely nauseous. They had to stop the process. It took the nurses several hours over two days to get her cells back into her.

One of the primary claims we'd made to the FDA when requesting approval for the Ceprate system was that because substantially less DMSO was needed to store purified stem cells, infusional toxicity was greatly reduced. This was about as good a test of that claim as it was possible to make.

This woman had a bad time of it. While we were both in isolation, Patty had spent some time with members of her family. They'd all flown to Seattle from New York. After the infusion, Patty met them again in the waiting room. "I know it's really hard on the patient," a relative said, "but it's just as hard on the family watching them suffer so much."

I wasn't suffering at that time at all. Because my cells had been concentrated before being frozen, very little DMSO was needed to protect them. The day after my transplant, a nurse came into my room and examined my chart. Finally she asked, "Are you on some sort of experimental protocol?"

"Yeah," I whispered, "why?"

"Well," she said, with a puzzled look on her face, "you've just been transplanted, and usually when that happens the whole room smells like DMSO, and I just don't smell it in here."

The results of this particular study seemed obvious.

The reinfusion began the most dangerous and difficult part of the transplant process. Even under ordinary circumstances a small number of patients, no more than 5 percent, simply don't engraft. For some reason their stem cells just don't take hold and start growing. Nobody knows why. When this happens in an autologous transplant, the backup apheresis product can be used; in an allogeneic transplant doctors take more bone marrow or blood from the donor and try another infusion. I wasn't worried about this, but I knew it was a possibility.

In my case there was the potential for another problem. Although stem cells repopulate the blood and immune system, other types of cells—so-called accessory cells—may be necessary to stimulate engraftment. During the B-cell depletion process, we worried about losing additional cells. That's an unavoidable consequence of running blood through two columns. While there was no reason to believe this would cause unexpected difficulties, the only way of knowing that for sure was to see what happened to me. Another study in progress.

The real evidence that my stem cells had engrafted and were producing new cells would be a white-blood-cell count. Until that cell factory inside my bone marrow was churning out new cells, there would be no way of knowing whether my transplant had been a success or a failure.

The problems I would face were infection, fever, and bleeding. All were relatively common, and all were potentially deadly.

For several days my body was basically a very large petri dish. It contained everything an invading organism needed to thrive, and no immune system to protect it. At that point almost any organism that took root could be very dangerous. Certain strains of bacteria replicate rapidly, feed off the nutrients in the bloodstream, and dump potentially lethal toxic chemicals. In a healthy body, white blood cells would surround and engulf the invading organism. When a cut gets infected, for example, it gets red and swollen. The stuff that comes out of it, pus, consists of white blood cells that have died fighting the infection. But I didn't have any white blood cells. I didn't have an army to fight for me. So something as ordinarily harmless as a cold sore, which is just a herpes infection, could spread throughout my body and, with no immune system to fight it, might be fatal.

A fever almost always indicates the presence of a foreign body in your system. It can mean that an infectious organism has gotten into your bloodstream and is multiplying and circulating. My fever started the day after I received my stem cells. And it got very bad very fast.

Many people don't know this, but fever is actually a strategic weapon of your body's defense system. Most invading organisms live and multiply at normal body temperature. Turning up the heat on these organisms is your body's attempt to destroy them by creating a hostile environment. That's the same reason doctors advise people who don't feel well to get into bed under the blankets or take hot baths. In the past, doctors artificially induced fevers with microbes to fight an invading organism.

But at times fevers can quickly grow out of control and cause permanent damage to organs or even death. Fevers are particularly dangerous in children because they can cause brain damage. While my doctors believed my fever was a neutropenic fever, meaning it was caused by a very low white cell count, they couldn't know that for sure. It might just as easily have been related to a serious systemic infection.

The biggest cause of death in transplant patients is infection. Three times each day, doctors had to take all my vital signs. I began to dread that, because every time they examined me my temperature was just a little higher—and I knew exactly what that meant.

My fever climbed to 102 degrees . . . 103 . . . The only weapons they had to kill invading organisms were antibiotics, but without knowing what was causing the fever, it was impossible to know which antibiotic would be effective. Whatever the cause of this fever, my doctors were extremely concerned. They treated me by trial and error. Every few days they would give me a new combination of antibiotics, then wait to see if it had any impact.

At 103 degrees they gave me the strongest, broadest spectrum antibiotics available. One of them, I never knew which one, caused a terrible allergic reaction. I broke out in a rash. This rash spread quickly over my entire torso and face. It itched, but the doctors were able to make that bearable.

The antibiotics were not working. My temperature continued to rise, up to 104 degrees. Nothing the doctors tried seemed to have the slightest impact. I spent most of the time in a morphine-induced haze, but even in that fog I knew I was seriously ill. I knew exactly what my fever meant, and I knew the consequences if they couldn't get it down.

Patty was frantic. She wanted so much to help, but there was nothing she could do except watch and wait. She sat by my bed day and night, asking questions, trying to figure out if there was anything else that might be done. Patty isn't someone who easily reaches out for help, but one night she called her close friend Phyllis Partridge, a nurse who lives in Massachusetts. "I just really need to hear your voice," she told her.

Phyllis was reassuring. "There's nothing else you can do," she said. "His temperature'll come down. You just watch."

As the days passed, my white cell count stayed at zero. Each morning the nurses took my count and posted it on a chart. It was zero, always zero. My immune system was not growing. Time was against me. The mature red blood cells and the platelets that had been circulating endlessly throughout my system were wearing out, they were dying, and my bone marrow was not capable of replacing them. So I began receiving transfusions of donated red blood cells and platelets almost daily.

Platelets are the first line of defense against bleeding. They seal open wounds. In a sense, platelets are the pothole fillers of the body. In addition to cuts in the skin, they find small breaks in the walls of the blood vessels and capillaries and plug them. They dump chemicals at the site of the wound, which form a clot, giving the torn tissue a chance to heal and grow back. Although most people aren't aware of it, platelets save your life several times every day by preventing you from bleeding to death. So they get used up pretty quickly. The transfusions were supposed to replace them. One night I developed a hacking cough. I also felt terribly congested, so I blew my nose. In the condition I was in, blowing my nose could have been fatal.

I felt a warm liquid flowing from my nose. Blood. Blood was pouring out of my nose, and I didn't have enough platelets left in my system to stop it. Blood covered my hospital gown. The nurses pushed back my head and packed my nose tightly with cotton. Eventually they managed to stop the bleeding.

It happened again the following night. This time the bleeding was worse, and more difficult to stop. I lost a substantial amount of blood. The doctors brought in ear, nose, and throat specialists to try to find the source of the bleeding. It was far back in my nose, too far for them to be able to seal it easily. So they just packed up my nose with wads of cotton to prevent me from bleeding to death.

These spontaneous nosebleeds worried the doctors even more. They had expected some bleeding, but this was a potentially very serious problem. They decided to do a CAT scan of my sinuses to see if they could locate the source of the problem. Climbing out of my bed and into a wheelchair, even with assistance, was really tough. I felt so strange as they wheeled me through the hospital corridors that I might just as well have been exploring the surface of Mars. During the CAT scan, the only thing I was required to do was lie still. That exhausted me.

Five days after the transplant, I had a high fever, my body was covered with a rash, I was bleeding, and I had no white blood cell counts. After six days, the same. Seven days, the same. As Patty sat with me one night, she reminded me on the scene from another favorite family movie, *Bull Durham*. In that film everything is going wrong for the team and the catcher (Kevin Costner) shakes his head and says to the coach, "We're dealing with a lot of shit here." I didn't have the strength to laugh, and my mouth was so raw it hurt to smile.

I remember lying in bed after finishing chemo and running my tongue over the roof of my mouth—and realizing it was gone. The skin that remained had sort of buckled—basically my mouth and digestive system was raw, exposed tissue. Every time I moved my lips or tried to swallow, an excruciating pain ripped through my mouth and throat. I couldn't eat or drink anything, not even water. Yet I never felt hungry. Basically, my digestive system had shut down.

The pain was intense and it never dissipated. It was like having surgery without anesthesia. Compared with that pain, nothing else seemed to matter. While I was aware of all my other medical problems—the fever, the rash, the bleeding, the lack of counts—they seemed so minor in comparison to the pain. They didn't even make me any more uncomfortable.

To combat the pain, I was put on a morphine drip. At first I didn't want to use narcotics, because morphine destroys one's sense of reality, but Sherri Bush pushed me hard. "There's no benefit to not taking them," she told me. "We don't give a prize at the end to people who get through it without them. And if you don't take them, you're going to be miserable." Still, I resisted it as much as possible, and I was very proud of the fact that I took less than half the amount needed by most patients.

I took morphine and Benadryl, and drifted in and out of consciousness for more than a week. There was no night and no day. I was aware when people came into the room, but their presence had no context in time. They could have been standing there for a minute or a day and I wouldn't have known the difference. They could have been there hours earlier and I would have no memory of seeing them, yet if they had not been there for days I might have thought they had just left. All my connections to time and space were severed. One day Jamie and I were in the middle of a normal conversation until I closed my eyes, and then we were together in a totally different situation. We were still in my hospital room, but different people were there with us. I was with Jamie in two different places, holding two conversations, and I would shuttle back and forth between them simply by opening and closing my eyes. That scared me. I was completely aware of what was going on, even intrigued by it. I'm living in two different worlds, I thought, but I also worried that my mind was slipping away.

I had no idea what I was saying. Another day Jamie and Ben and Patty were there and I was almost asleep. I could hear their voices. And suddenly I asked Jamie, "Son, do you like women?"

To which he replied, "Sure, Dad."

"Good talk, son," I said. That was a line from another one of our favorite films, *National Lampoon's Vacation*. Several minutes later, back in my stupor, I babbled to whomever I was speaking with in my mind, "Jamie's out with those girls again. I don't know what to do with that kid."

I dreamed wonderful dreams. I dreamed of being on the waterfront; oddly enough, not on a boat, but in a harbor. It was a safe place to be. It was a very vivid dream, and while I don't remember the details of it, I remember the feelings quite well. And I had bizarre dreams about being in a spaceship and hurtling through space.

I hallucinated. While I was in isolation, Patty had taken a three-day art class to keep her mind occupied. In that class she made three mixed-media collages. She hung those collages on a bulletin board in front of the bed. When I opened my eyes, I looked right at them.

These collages were completely abstract, completely free-form. But as I stared at them, I began to see things in them that I hadn't seen before. At times the collages came alive for me. They actually moved. I would see beach scenes in which the waves would actually break and roll into my room. The first time that happened, I thought, *Oh, there's a very distinct*

pattern there that I must've just overlooked before. How could I have missed that?
One of the collages, I became convinced, depicted a garden party, and
among the guests were Bill and Hillary Clinton. In this collage I saw Bill
Clinton standing by a car and smiling. The car was so crystal clear, so distinct, that I was amazed I had never seen it before. I couldn't wait to tell
Patty that I had figured out what she was trying to do with it. When I told
her, she was very solicitous. "That's nice, dear," she said. "Are the Clintons
still there?"

Of course, when I looked at it the next day I saw something entirely
different. The car, and the Clintons, had disappeared.

At times I was delirious. One afternoon, or late one night, I decided
I couldn't take any more. I started looking around the room, as if searching for something. I told Patty, "I'm going home, babes, I've had it, I'm
going home." Apparently I tried to get out of bed and collapsed back into
it, admitting, "Uh, well, guess I can't."

But I remember people coming in each day to take my blood count.
I remember their disappointment when I had no count. I remember the
straight line on the wall chart. And I remember being aware that my life
was completely in the hands of the doctors. Guys, I thought, I sure hope
you're doing everything you can, because there isn't much I'm going to
be able to do to help. All I had to offer was my basic survival instinct. I
knew that eventually I was going to come out of this tunnel. I just didn't
know when.

I waited every day for the slightest evidence that my stem cells had
started to engraft. Just a flicker of a count was all I needed. Just something
to give me hope. Just a hint that everything that had been done at CellPro
and everything I had been through was worth it.

Patty was there every day, staying as long as the staff would allow.
When I drifted into sleep, she would be there, and when I woke, she
would still be there, rubbing my feet with lotion.

After eight days I still showed no counts. Nine days. Whatever was
happening inside my body was happening very slowly. If anything at all was
happening. There was little physical evidence that I was improving. When
my fever reached 104 degrees, the doctors became extremely concerned. At
105 degrees they would have considered packing me in ice to cool down
my organs. The cotton packing had stopped the nosebleeds, but my platelet
count remained dangerously low. The normal platelet count is about
200,000 to 500,000 per cubic microliter. Below 20,000 there is a serious

danger of spontaneous bleeding. I was below that number. So I continued to receive transfusions. The rash would not go away. Sherri kept telling me I looked very well for someone who had just been through a major transplant. I grunted, the closest I could come to a laugh. If I was looking well, I couldn't even imagine what people who were not looking as well looked like.

I measured the length of the days in pain. My mouth and digestive system were still a wreck. Good days were those when I managed to sleep several hours.

My existence at CellPro seemed like a life lived on another planet. The people of CellPro had been completely supportive. When I entered the hospital, they'd sent me a paper banner about ten feet long that everyone had signed. Next to a drawing of a sailboat, someone had written, "Wishing you a breeze to recovery." Referring to the fact that I'd used antibodies taken from a mouse, Joe Tarnowski wondered, "Do you have any taste for kibble or cheese since the transplant? Hurry back!" In my former life, Larry Culver and I had eaten lunch together at Cardeli's Café every day, causing Mark Lodge to report, "Larry has been spotted at Cardeli's having conversations with an imaginary friend. Hurry back, we all miss you." Every message was upbeat, optimistic: "It's too quiet in our corner with you gone," wrote Joanne Reiter. "I miss hearing your voice and laughter at the other end of the hall. Get well soon."

And maybe someone I hadn't even met summed up best what CellPro was all about. "Hi, I'm new here," she wrote, "but I think this place is great. Thanks for making it so. Hurry back."

A lot of people sent me presents. I unwrapped only a few of them, I was in no condition to enjoy anything. But all of this combined to make me feel that CellPro was my home and that these people were my family. I missed this family very much.

Even in my half-conscious state, by the tenth day I was getting very anxious. Sherri insisted everything was proceeding quite normally. "Keep it in perspective," she insisted. "Look how hard it was to kill your immune system. Ten days isn't a very long time for it to start growing again." Sherri was right, of course; I was living through an amazing process, the rebirth of one of nature's most complicated systems. It had taken tens of millions of years for the immune system to develop into a weapon capable of protecting the body against almost any invaders, and I was worried because it hadn't taken root in my bones in ten days.

I believed Sherri completely, but I wouldn't feel comfortable until I had a count.

On the morning of the eleventh day, a nurse took my counts. "Well," he said with satisfaction, "looks like we're getting a flicker of a white count." It wasn't much, less than the whisper of a breath, but it was real. It was the heartbeat of my immune system. The normal white count at UW was between 5,000 and 10,000 per cubic microliter; I had fewer than 100 per cubic microliter, too few to count accurately. But it was a beginning.

The next day I had real counts. The word spread rapidly throughout the hospital wing. People kept sticking their head in my door to congratulate me. "Counts today!" It seemed as if everybody wanted to talk about it. Patty could barely contain herself. The boys came to see me. It meant that I was safely reaching the end of the tunnel. My count was still very low, barely 100. The assembly line in my bone marrow was slowly cranking up. It was as if every cell had to be made by hand. But that number was vitally important, because at that level the immune system can fight life-threatening infections.

My fever broke. Finally my temperature started to decline. The rash began abating. I started to come out of the mist in which I had been existing. When I looked at Patty's collage, the Clintons were no longer enjoying the garden party. The doctors finally took me off morphine, but gave me a sedative to help me sleep at night.

The next morning I woke up and heard someone moving around my room. I assumed it was a nurse. But when I opened my eyes, I was absolutely shocked to see a delivery woman from Airborne Express holding her clipboard and a box. And she asked me to sign for the box. *This is interesting,* I thought. *I must still be hallucinating.*

But she looked very real, and she again insisted I sign for this box. I began to realize it was not a dream. I signed for it. She set the box on the chair and left wordlessly. A few minutes later a nurse came into the room and saw the box and did a double take. "What's that?" she asked.

"I don't know," I told her. "A delivery lady just brought it."

A few hours later I finally got enough energy to see what was in that box. The CellPro people in Europe had sent me six bottles of brut champagne. It was Veuve Clicquot, my absolute favorite. I left it on the chair where I could see it. Drinking that champagne became my goal. I knew that when my throat felt well enough to allow me to enjoy that champagne, I would be all the way back.

That day my count reached 500. That's considered the threshold level for daily living. The assembly line was finally churning out white cells. What had been a gradual rise on my chart was turning into a slope. I was starting to feel a lot better. The fever had just about disappeared; my doctors never did discover its cause. Even my mouth started healing.

But I still couldn't eat. My taste buds had been destroyed, so food had no taste at all. I couldn't even keep water in my stomach. My gastrointestinal tract had been devastated and had to be rebuilt. I was being "fed" intravenously through my Hickman catheter. One night Jamie came to visit, and I decided I could eat a Popsicle. Just sucking on it soothed the sores in my mouth, but I couldn't even keep flavored ice down. That was so frustrating.

I wanted my life back. I wanted to go home. I wanted to get back to work. But after lying in bed for almost a month as my body was turned inside-out, I had no strength. Standing up was about the extent of my physical activity. And until I started eating I wouldn't be able to regain that strength. Ollie knew that I was anxious to leave the hospital. "What do you think?" he asked me one day. "What's your game plan here?"

I remember my exact words. "I just want to get the fuck out of here."

"Well," he said, "then you need to start eating. We need to see that you can keep stuff down, because that's going to be the next big hurdle." We agreed that when I was able to keep one thousand calories in my stomach I could go home. One thousand calories. Once that had seemed like nothing. Before I'd gotten sick, when I had been trying to lose weight, it seemed as though everything I ate was more than a thousand calories. But at that moment, lying in my hospital bed, one thousand calories might just as well have been a chocolate Mount Everest.

There was a small deck at the end of the hall, and sometimes in the late afternoon I would get into my wheelchair and Patty would wheel me out there. Just sitting a few feet outside the hospital was victory. Then one day Patty decided, "Let's really go outside. Let's go downstairs."

Downstairs? Outside? We were going to escape. Patty and I had shared a lifetime of adventures, but never anything like this. Adjoining the hospital is a lawn that leads down to a small pond. She wheeled me down to the water. It was as if I were a blind man who'd suddenly regained his sight. Everything seemed new and fresh. The birds were singing loudly, and the leaves were rustling. As we got closer to the pond, I could see insects skimming the surface. I could smell the freedom in the air. I was alive.

Applesauce was the first thing I was able to keep down. It had no taste but I didn't care. I kept it down. Then I was able to eat just a small amount of yogurt. And finally I discovered Ensure. Ensure is a high-calorie drink made especially for senior citizens, but for me it was as delicious as the champagne. The day I was able to drink one thousand calories of Ensure Plus and keep it in my stomach was one of the most satisfying days of my life.

When I told Ollie, he smiled and sort of shrugged. "Then I guess there's no reason for you to stay anymore," he said.

When Patty arrived that day, I said, "Let's go."

She didn't quite understand what I was talking about. "What do you mean?"

"Let's go," I repeated. "We're going home."

She hadn't expected it, but she was thrilled. It was important to me that I walk out of the hospital, but I told my nurses that I wanted a wheelchair. "I've got all these presents," I explained, "and I'm going to need a way to carry them." I also knew I would need the chair to lean on. I put on my own clothes for the first time in weeks. We loaded Patty's collages, the champagne, and several other gifts I'd received on the seat of the wheelchair and I walked slowly to the elevator and out of the University of Washington Hospital. It was one of the longest walks of my life. It had taken months. In addition to all of those presents, I carried with me a brand-new immune system. Happy rebirthday to me.

When we got to the car, Patty opened the passenger door for me and I climbed inside. She put the collages and champagne in the trunk and took the wheelchair back to the hospital. As I sat there waiting for her, an overwhelming feeling of satisfaction flowed through my body, maybe hitching a ride on my new white cells. *I made it*, I thought. *I made it.*

Patty drove us home. Home had been many different places in our life together, and for me it had never really mattered where it was located. Home was being with my family, which included Fat Ass the cat. Although Jamie had returned to college at Worcester Polytech and Ben had left for Bates, home had never been more welcoming. I found great comfort in the familiarity of my own environment.

At night Patty would sit with me on the bed and read aloud. When we lived in Belgium she had become interested in the World War I battle at Verdun. One million soldiers had died there, among them one of her favorite German expressionists, Franz Marc. The first book she read to me was Georges Blond's *Verdun*. It read like a novel; as she read, we could hear the thunder of the endless bombardment, and feel the terror of the soldiers as the officers blew their whistles ordering them over the top to near certain death. On those nights we shared their horror, and it was easy to forget that I was struggling to regain my own life.

When that book was done, we began reading Christopher Martin's *Battle of the Somme*. We both became extremely interested in World War I, and we read as much as we could, then began renting videos to see this tragedy for ourselves. We saw the defoliation of Europe, with the great forests completely leveled. We saw the bodies of the soldiers, and we saw horses struggling desperately to pull wagons through fields of mud. Neither of us will ever forget the vacant stares on the faces of those young soldiers.

Who would have suspected this terrible war would have been such wonderful therapy for me.

When I'd left the hospital, Ollie had warned me not to try to do too much, too quickly. "The best thing you can do is walk," he explained, "but you have to be real careful. If you overdo it, you can really hurt yourself."

Hurt myself walking? I didn't think that was likely. I'd been a serious runner for years. When we moved into the area, I'd staked out a three-and-a-half-mile course through the neighborhood and ran it often. Within days of coming home, I tried walking that course. I walked around the block, at most a half-mile, and doubted I could make it home. I was sweating profusely, I got very dizzy, I had absolutely no stamina. I had to stop several times and rest. For someone who had always taken pride in being physically fit, I was appalled at this evidence of my body's betrayal. When I finally got back to the house, I collapsed onto my bed and lay there with my eyes wide open, staring at the ceiling. Nicole and the Rick Project team and Ollie Press and his staff had given me back the tools for life, but it was up to me to use them.

I had been in such a hurry to get home because being home represented a return to my normal life. But actually being home forced me finally to accept the fact that I wasn't the same person I'd been weeks earlier. Even though I was in familiar surroundings, I could no longer do the same things I'd always done so easily. Successfully walking up the stairs was a major event. It took me a long time to accept these limitations.

I walked every day. There was a shallow stream about a mile down a long hill from the house, and that time of year the coho salmon were running up it. The annual salmon run is one of the miracles of nature. These beautiful bright red fish fight their way upstream against the current, against the rapids, leaping over falls to return to the area in which they hatched to lay their eggs. There is something magnificent and terribly sad about it. They were desperately trying to get home to give life, and when that was done they would die.

This was near the end of the run, and these fish were well spent. There was a small pool near a footbridge, and as they rested there to gather strength for the last part of their journey, I would stand there for hours watching them. They still had some fight in them, but they were starting to pair up. In an afternoon I would see fifty or more fish. After the coho salmon came an occasional king salmon. King salmon are big fish, weighing twenty-five to thirty pounds. I had absolutely no desire to drop a line, I just thrilled to watch them. At times I thought I should put up a little yellow road sign reading NATURE AT WORK. This was an event most people experience only on cable television, but I was able to walk down the hill from my house to see it. I loved it. Getting down to the creek, then back up the hill, became my goal.

In addition to Patty, I had the company of old Fat Ass. That cat was amazing. Our relationship had always been one of mutual toleration. But while I was recovering, she sensed my need for companionship and would come and sit close to me, which was completely out of character.

Ollie kept me on a very short string. For the first two weeks I had to return to the hospital for blood tests almost every other day. Back and forth, back and forth, Patty referred to it as "the Curly Shuffle"—in tribute to the Three Stooges' Curly Howard's habit of repeating the same action over and over. Then it was once every three days, then four days. . . . In late September, Ollie decided it was time to do a complete follow-up. This was gut-check time for me. Time to find out if the whole process had really worked. It was very strange, returning to the cancer center as an outpatient. Sherri Bush and Dr. Liu ran the tests. They did a CT scan of my lymph nodes, a complete blood workup, and a bone-marrow biopsy. The fact that my immune system had been regrown was proof that the stem cells had engrafted and were pumping out new cells; the question was whether the tumor cells had been eliminated or also had taken root.

After completing the tests, I had to wait almost a week for the results. Basically, I held my breath the whole time. My fantasies played tag with my fears. I anticipated good news, but I tried to prepare for devastating news. Patty and I returned to the cancer center the following week to learn the results. It seemed as if just a few days earlier I'd been waiting in Dr. Linville's office, but that had been a lifetime ago. As we waited to be called, I casually leafed through some magazines, but I really couldn't focus on anything. I was about to get the first real verification that my death sentence had been commuted.

Ollie was smiling when we walked in, a smile that told me everything I needed to know. My system was clean. There was no evidence of disease. My spirit soared at that news, but I didn't get excited, I didn't even raise my voice. And I absolutely refused to use the word *cure*. I am not one to tempt fate. But if the enemy was still alive in my body, it was deeply hidden.

For two months I returned to UW once a week for a checkup, then graduated to once a month. The final test for me was that my hair started to grow again. No matter how often I had been reassured that my hair would grow back pretty much as it had been before chemotherapy, it was a great relief when I looked in the mirror and saw a field of dark fuzz. When I had to start shaving again, I asked Patty if she wanted me to regrow my mustache. It was her decision. "Don't shave it," she decided. "Just let it grow."

One perfect September morning a breeze caught my imagination. I was feeling better, the sun was shining brightly, and the wind was persistent. It seemed like a good day for a sail. "It's time, Patty," I said, "let's go to the boat."

I had to talk her into it. It had been many years since Patty and I had taken a sail alone. She was always warning me to go slower, take my time, don't overdo it, and remember what Ollie had told me. Without the boys to raise and lower the anchor and the sails, I knew she was concerned I would try to do too much. But it was such a perfect day for a sail.

It was tough. We struggled to get up the sails, it was difficult for me to raise and lower the anchor. Patty worried through the day. But together we managed a lovely sail up to a place called Port Ludlow. Whatever strength it took out of me, I got back many times in satisfaction. Day by day I was getting my life back.

About a month after leaving the hospital, I learned from Larry Culver that CellPro's managers were meeting off-site to discuss marketing plans for the Ceprate system product launch. Although we still did not have final FDA approval, we expected to receive it almost any day, and we had to be prepared to get going immediately. I told Larry I was going to try to make this meeting, but asked him not to tell anyone. Just in case.

Joe Tarnowski had just arrived when I pulled into the parking lot. "You look great," he said, then added, "a whole lot better than I'd heard."

"Well," I said, "you didn't see the worst of it."

Joe was thrilled. "This'll be a real boost for everybody seeing you walk in there."

As I walked into the meeting room, the first person I saw was Nicole, sitting at the table reading some documents. She didn't notice me right away. It would be impossible to describe the spectrum of feelings that I felt welling up inside. What do you say to the person who did so much to save your life? "Thank you" doesn't begin to express the way I felt. So I just put my arms around her and we hugged each other.

And then we started arguing. I'd thought I would be much too weak to stay for the entire meeting. But when I got there, the adrenaline started flowing and I forgot to be tired. I expected to act as an observer, just to listen. Well, that lasted about five minutes. Then I was right back in the thick of the fight.

After completing plans for the product launch, we began discussing other projects—among them the B-cell depletion system. In my absence the Rick Project had become known as BCD, B-cell depletion. And I was stunned to find out that Nicole and her team were strongly against taking the process into clinical trials. I was sitting there, alive, cancer-free, and I believed it was in part because of this device. I wanted other people to have the same chance I got. I was adamant that we get this thing into the clinic as quickly as possible.

"We're not ready," Nicole insisted. "We've got to go back and re-engineer it. It just isn't user-friendly the way it is now."

I was so upset I lashed out at her. "Look," I told her, "it worked for me. We need to get it out there. Ollie needs it, he's asked for it, we really need to move it ahead," I insisted. So much for sitting back as an observer.

"I'm telling you, Rick, we can't do it," she repeated. "It's not possible to standardize the blocking reagents."

To me, she looked as if she might start crying. I thought to myself, *What the hell am I doing?* But I felt just as passionately as she did. I believed this device had helped save my life, and whatever it took, even if it was held together by chewing gum, it was going to be out there.

This was an argument I was going to win. I insisted we immediately take steps to turn it into a product.

I started going into the office for a few hours each day. It seemed like just the right amount of time: Patty thought I was working too hard; I didn't think I was working hard enough.

At work, Nicole and I battled often about the BCD project. Ollie Press was as anxious as I was to get it into clinical trials, where it would be readily available to his patients. He asked me to put together another device for a second compassionate use. Although Nicole continued to insist that the device was still not ready, the team built another column. I was almost as excited when this second patient was transplanted as I was when my cells were purged. This was to be the proof that the B-cell depletion system worked.

But the patient died. Her death resulted from a combination of factors. The patient did not mobilize well, so we had a smaller population of cells with which to work. And then, for a reason no one could explain, somewhere in the system we lost a lot of stem cells. The process that had worked perfectly for me just didn't work very well for her. It was just an anomaly, I decided. But I realized how little we knew about the system. We began backing up, doing all the experiments we should have done prior to my transplant. Backfilling, we called it.

I never knew this patient's name. I tried not to identify too strongly with her. Death was not unusual in this business. But there were odd moments when I found myself thinking about this patient and feeling great frustration at our inability to save her life. It wasn't our fault, I knew. There were just too many variables. I don't remember ever comparing her case to mine. And outwardly at least, I acted as if this were business as usual.

But I knew it wasn't. Not anymore. It was as if I were holding on to a life preserver and I couldn't rescue everyone.

In December, Ollie Press gave a major address at the annual meeting of the American Society of Hematology in Orlando, Florida. Before his speech he had asked Nicole for some data about the system. "You're not going to put in Rick's data, are you?" she asked with great apprehension.

"No," Ollie said, "but I do intend to talk about it."

Nicole and Sharon were sitting in the audience for Ollie's speech. Tumor-cell purging is still very controversial, he explained, but if you have the ability to remove the tumor cells, why would you put them back in? "CellPro has been working on a new system," he continued. "We were able to get four logs of depletion with this device. It has been used with one patient."

Nicole looked at Sharon and shook her head. "Oh, no."

As we all filed out of that meeting, I caught up with them. Neither one of them was very happy. "Well, ladies," I said, "here we go. Looks like Ollie's already sold it for us." The floodgates had opened.

I had returned to work full-time—against Ollie's advice—in November. My mustache was growing back, but I was still wearing my hairpiece. I had stopped wearing a hat. I continued returning to the hospital for tests every month—then waiting anxiously until Sherri called me to confirm that all the tests for the presence of tumors were negative. But I was careful never to use the word *cure*.

Rather than praying for a miracle, I had put my faith in science. But so many people who learned about my illness told me, "I just want you to know that I'm praying for you." These were people of all faiths. I gained enormous comfort from the fact that these people thought enough of me to do what they believed was the most important thing they could do. I know I was the recipient of many prayers. And sometimes even I wonder at the incredible series of coincidences that found me at the only place in the world where my life could have been saved. Maybe it was just luck, or fate. Maybe.

The fact that I felt so strongly that CellPro's technology had helped save my life made me even angrier about the patent lawsuit. There was no other system capable of doing what we had done, and this lawsuit put its very existence in jeopardy.

I returned to work at CellPro just in time to prepare for the second trial. It seemed appropriate that I would lead the company into legal battle. I mean, who better? More than ever, I was determined that this technology would continue to be developed and become available to patients who needed it.

My philosophy in business has always been that competition is very positive. It drives progress. It forces the competitors to be better. This was a marketplace in which everyone involved, from CellPro and Baxter to the patients, benefited from the competition. We were developing a new area of therapeutic medicine, and there was plenty of room for two or more players. At least that's what we thought. Apparently our competitors believed differently. They wanted to drive a stake through our corporate heart. They wanted the market to themselves. Never before had this cliché been more appropriate: they wanted to bleed us dry.

Judge McKelvie had scheduled the second trial to begin on October 15. Prior to the beginning of that trial, based on the Supreme Court decision in the *Markman* case, which gave judges the power to determine the meaning of patent claims, McKelvie stated that his interpretation of the Civin patents in the first trial had been "in error." He ruled that the instructions he had given the jury had been incorrect. So before the second trial began, he was going to decide what the patent claims meant. He was going to reinterpret them.

Baxter submitted its interpretation of the meaning of the patents. Naturally, this interpretation was extremely favorable to their case. While the court acknowledged that the claims construction proposed by Baxter was "highly unorthodox," McKelvie accepted them. Basically, he decided that the claims meant what Baxter's attorneys said they meant.

It was astonishing. To me, this was like letting the home team rewrite the rule book in the middle of the game. As we proceeded, it was becoming clear we were in uncharted territory. I had been told by our attorneys that it was almost without precedent for a judge to overturn such a powerful jury verdict. The jury in the first trial had ruled unanimously in our favor on every one of 103 questions. There was no ambiguity in their verdict, no indecision. And as *Markman* had only recently been decided, there were few cases in which judges had rewritten patent claims, and to my knowledge there was no case in which the patent claim had been completely rewritten. At one point we even filed a mandamus, a request asking the appellate court to intervene in the case immediately.

McKelvie was rewriting history, expanding the Johns Hopkins patents far beyond what had been originally granted to them by the patent office. *Rewrite, reinterpret, reconstitute,* all nice words to say the same thing: Judge McKelvie changed completely the meaning of these claims. In the first claim, I believe only five of the original 137 words remained.

That first claim, of the 204 patent, covered ownership of the My-10 antibody discovered by Curt Civin. Civin made a monoclonal antibody, My-10, that recognized stem cells. But My-10 had no commercial value because it couldn't be biotinated and it did not react with primates, with monkeys, so it couldn't be used in animal tests. Civin tried to make another antibody, one that could be biotinated and did react with primates, but he failed. In fact, using his own methods he was never able to make another antibody in his laboratory.

Admittedly he got there first; he made one hybridoma capable of producing, on a continuous basis, the My-10 antibody. But if he couldn't make another one, his patent certainly didn't teach anyone else how to make stem-cell—or CD-34, as they were officially designated—antibodies. So, under patent law, he did not enable it.

When an inventor files for a patent, examiners from the patent office determine whether the invention does what the patent claims it does. Often that leads to significant changes being made in the claim. The patent office keeps a good record of this process. Apparently, Civin had made very broad claims for his antibody. But as the brokerage firm Auerbach Pollack & Richardson, Inc., one of several companies that carefully followed CellPro, reported to its clients:

> *[T]he Court of Appeals will probably reverse Judge McKelvie's decision on the 204 patent [because] the patent claims ownership of a potentially huge and undefined collection of antibodies, even though those antibodies are not enabled in any way by the patent. If the claim had been limited to the use of antibodies to fish out stem cells, the claim might have been narrow enough, but this claim, as interpreted by the court, covers any antibody that could ever stick to a CD-34 molecule that also happens to have a My-10 epitope. There are potentially hundreds of thousands of antibodies that satisfy this criterion, and that is a lot of antibodies to take out of the public domain in a single patent.*

The brokerage house pointed out that the patent examiner had recognized that only one antibody, My-10, had been enabled, and therefore almost all the other claims were clearly limited to My-10.

Not according to Judge McKelvie. Civin had discovered a single antibody that binds to stem cells. In his patent he "enabled" it, meaning he explained how it could be made by someone else "schooled in the art," someone who knew what he was doing. But McKelvie granted Civin the rights to every antibody that binds to a stem cell, even though those antibodies could not possibly be made using the information in Civin's patent claim. Civin didn't teach anyone how to make a stem-cell antibody. Unfortunately, included in that vast grant of rights was our 12.8 antibody.

In the 680 patent, the second patent, Civin claimed rights to any stem-cell suspension "substantially free" of other cells. McKelvie interpreted "substantially free" to mean 90 percent pure. The Ceprate system was never intended to produce a 90-percent-pure suspension. A purity level that high could be obtained only by sacrificing yield, and in the laboratory when our system produced a 90-percent purity level we often didn't get enough stem cells for a successful transplant. We also found that lower purity levels seemed to engraft better than 90 percent and above. But Judge McKelvie decided that because the Ceprate system theoretically *could* be used to create a suspension 90-percent pure, we had infringed this newly interpreted claim.

That is the legal basis on which we were forced to fight. But it was only the beginning. This wasn't just an uphill fight, this was an up–Mount Everest fight. Every ruling Judge McKelvie made seemed to favor our opponents. For example, Judge McKelvie scheduled the second trial to begin in mid-October. After the first trial, our attorneys had spoken to jurors and found that I had been among our most credible witnesses. The jury had liked me and believed me. Obviously I was going to be an important witness in the second trial. But I had just gone through a major medical ordeal, and I was barely back on my feet. Ollie did not want me to have to deal with too much stress at that point in my recovery, so Patty was going to return to Delaware with me. In addition, Lyon & Lyon had had only four months to prepare a completely new defense strategy based on McKelvie's new patent interpretations.

With all those factors to be considered, Coe Bloomberg had requested an extension. McKelvie turned us down.

Lyon & Lyon did an incredible job preparing for the second trial. In response to Judge McKelvie's reinterpretation of these patent claims, they worked day and night to put together our new defense. On the very last day possible, they submitted to the court ten new reports from experts in the field. In addition to those reports, they had also put together a compelling case based on the new claim construction. At one point, Coe Bloomberg had told me with satisfaction, "You know, Rick, it's a strange thing. But after everything that's happened, I think our case is stronger now than it was before."

Josh Green, who had doubts about our original case, agreed. When Lyon & Lyon presented our case in pretrial hearings, Don Ware, the attorney representing Baxter, BD, and Johns Hopkins, had probably figured out that we had a very good case. Actually, I think he was shocked. So he asked Judge McKelvie for more time to prepare—and Judge McKelvie gave it to him! The trial was postponed until the following March. The judge wouldn't give us the time we needed to prepare our case, but he gave it to them.

Not only were we climbing Mount Everest, we had to climb with our hands tied behind our backs.

Perhaps Judge McKelvie had been surprised at the strength of the defense Lyon & Lyon had mounted in such a brief period of time, or maybe he really believed he was proceeding impartially, but in a series of decisions over the next few months he took away from us any possibility of a second trial in front of a jury. It was an extraordinary display of judicial arrogance. Contrary to common belief, especially my own, in certain civil cases there is nothing that guarantees the right to a jury trial.

The purpose of a trial is to establish the facts in a case. If the important facts are not in dispute, meaning that the facts are so obvious that all reasonable people would agree to them, a judge has the right to make definitive rulings from the bench. Essentially, he can decide that, based on the established facts, there can be only one possible legal outcome. Known as "judgments as a matter of law" or "summary judgments," they give enormous power to a judge. Enormous. For example, if I were videotaped driving my car through a red light and hitting another car, no matter how many excuses I made, the facts could not be disputed. I drove through the light, I hit the other car. It wouldn't be necessary to have a trial to determine those facts. So the law allows the judge to rule on the case from the bench—but only when there is no dispute about the material facts of the case.

If there is the slightest dispute about the facts, you are entitled to a trial. When one party in a lawsuit requests a ruling from the bench, which

Baxter did, by law the judge has to give the other party the greatest benefit of the doubt—meaning that if you've got any case at all, you're entitled to a trial.

Beginning in October, through a series of judgments as a matter of
law and summary judgments, one by one Judge McKelvie began eliminating all of our defenses. It seemed as if we had endless hearings before him
and we lost them all. I could see Coe Bloomberg starting to get discouraged. At one point he warned me, "It's going to get a lot worse before it
gets better." As it turned out, Coe was very wrong about that—it never
got better. With every ruling the situation continued to get worse and
worse.

Once, when I was in college, I went skin diving with a friend off the
northern California coast. We intended to explore a small underwater rise
about twenty feet deep, but to get out there we had to swim over deep
water. When we reached the site, we anchored our floats and got ready to
dive. As I cleared my mask, I looked down and saw what appeared to be
about a fifteen-foot-long white shark. I'm sure my eyes just bulged. I told
my diving buddy, "I saw something down there. It looked like a big
shark." That was enough for him. We hauled up our anchor line and
began swimming for the beach.

I was terrified. I knew that sharks were attracted to several different
sensory stimuli. They respond to blood in the water, which they can sense
for many miles, and they respond to a thrashing on the surface. To a
shark, those things mean dinnertime. For us, the only way to get out of
there was to kick our way back to shore, which meant thrashing on the
surface all the way to the beach. We had no choice. As I began swimming
back to the beach, I could just feel that shark coming after me. With each
stroke I wondered when it was going to attack. It seemed inevitable.

I remembered that feeling when Judge McKelvie began making his
rulings against us. Once again I felt as if something terrible were closing
in on me. The result also seemed inevitable. Long before Judge McKelvie
eliminated our final defense, Bloomberg had said flatly, "Look, Rick, we're
dead with this judge. We've just got to look beyond that for the appeal."

This entire process stunned me. Throughout my life I have had great
faith in the American judicial system. Almost everything I experienced in
the first trial had reinforced that belief. So it was difficult for me to accept
that the same system bestowed on one man the power to dismiss the
results of that trial and award victory to the defeated party. Initially I was

simply incredulous; I didn't believe that was actually happening. I was naïve, of course, I believed that justice and fair play mattered. But once I realized that we had little recourse, I became very frustrated, and very angry.

But there was really no place for me to turn for relief. To my great dismay, I was learning that the law was not some sagacious code by which we have all consented to live; rather it was simply whatever Judge Roderick McKelvie said it was. Based on my experience, I have lost my faith in the legal system.

The shark was catching up to me.

PART THREE

THE VERDICT

"I STRONGLY BELIEVE THAT IF THE CELLPRO
DEVICE WERE FOR ANY REASON TO BECOME
UNAVAILABLE FOR MY USE, MY RESEARCH PUR-
SUITS WOULD SUFFER A SERIOUS SETBACK
AND THE INTERESTS OF MY PATIENTS WOULD
BE COMPROMISED—FATALLY, IN SOME CASES."

—*Richard Burt, M.D., Northwestern Memorial Hospital*

11

EVEN AS JUDGE MCKELVIE CONTINUED TO rule against us on just about every point, I refused to believe that CellPro could be driven out of business. That just wasn't possible. Rational people would not allow that to happen. If we went out of business, people would die. It really was that simple. Baxter's Isolex system was still in development; it remained years behind our Ceprate system. I didn't know how it would happen, but I knew that somehow a compromise had to be found.

In late February 1996, while I was struggling through my second round of chemotherapy, the FDA finally convened an advisory panel to determine whether or not the Ceprate system should receive government approval. Without this approval we were out of business no matter what happened in McKelvie's courtroom.

Three to four times a year the FDA convenes a panel of experts in a specific area to review data and advise it on new product approvals and regulatory policy. By law the FDA does not have to follow the recommendation of the panel, but in fact it usually does. A positive recommendation from an advisory panel just about guarantees FDA approval, and conversely if the panel turns down an application, it is extremely unlikely it will ever be approved.

The Biological Response Modifiers Advisory Panel was convened in Bethesda, Maryland, on February 28, 1996. The panel consisted of about twenty doctors and scientists, some of them transplanters who had used stem cells. Service on an FDA panel is voluntary and prestigious. Rather than in a courtroom, the meeting was held in a hotel ballroom. We took ten people to the meeting, and we were seated on one side of a U-shaped table, the panelists and FDA representatives were seated on the other. I wore my hat throughout this hearing, having already lost my hair. It was physically very difficult for me to be there, but mentally it would have been impossible for me not to be there.

This was the first time a stem-cell therapy product had been reviewed by an FDA panel, so there was considerable interest in these proceedings. More than one hundred people attended, ranging from stock analysts to our competitors.

The FDA was very concerned about the safety of the cell-separation technology. When we selected for stem cells, we eliminated other cells— and some of those cells might be necessary for long-term engraftment. Nobody really knew. Generally the FDA tells the panel what questions it wants answered, and these are usually very leading questions. Sometimes it's possible to tell simply by the questions which way the FDA wants the panel to vote. When I read these questions, my heart just sank. It seemed apparent to me that the message the FDA was sending to the panel was that it had grave doubts about the need to approve any product in this area.

The FDA asked the panel to vote on several questions: Would you consider this product safe? Is it safe and efficacious? Do the safety concerns outweigh the benefits? Basically, do you think this thing ought to be approved?

The debate continued through the morning. At times I could see it going against us. While transplanters generally believed purging was beneficial, we had no statistical evidence to prove that. We certainly could not show that it increased long-term survival. A few panelists even asked why doctors would want to use this device.

At CellPro, about twenty people had squeezed into a small conference room to listen to the hearing on a telephone speaker. They stayed there for three hours as the debate continued, offering responses to both the questions and answers. Comments ranged from "Why's he asking a question like that? He doesn't understand at all," to "Uh-oh, we're going down in flames here."

During the session the panel was scheduled to take three votes. Three votes that probably would determine the future of CellPro and perhaps the entire stem-cell separation technology. As the panel prepared to take the first vote, I had absolutely no idea which way they were going to decide. The questions had been very complicated. I was sitting next to our Director of Diagnostics Development, Amy Ross. "Here we go," she whispered as the voting started. She grabbed my arm and held on tightly.

The first vote concerned the clinical benefits of reducing infusional toxicity. A yes vote was a vote in our favor. The chairman asked all in favor to raise their hands. The room was large and crowded, so it was difficult to determine quickly how many hands were raised. But as I scanned the panel, it was obvious that many members were not raising their hands.

Then the chairman asked those against to raise their hands. That was the vote against us. It seemed like an awful lot of hands shot up. *Oh, man,* I thought. I quickly began counting the hands held high in the air.

At CellPro the boardroom was absolutely silent as everyone waited for the chairman to announce the result. The futures of a lot of people were at stake in these votes. Finally the chairman announced that on this first question—a majority voted yes—that the sponsor (CellPro) had demonstrated clinically significant benefits. Amy released her grip. I took a deep breath. But the fact that the result had been split concerned me tremendously. To me, this was the easiest question. If panel members voted against us on this one, it would be much tougher to get a majority vote on the really complicated issues.

After more debate the panel took the second vote. Did the Ceprate system actually do any good? Once again, Amy gripped my arm. In favor? Hands went up. I quickly tried to get a count, but lost track and had to start again. By the time I finished, the hands had been put down. Opposed? Again, hands were raised. This was also a split vote. There was more discussion. It was agonizing.

Finally it was time for the vote that would determine whether we stayed in business, whether the entire field of cell selection was beneficial or just technological snake oil. The question was: "Based on the data presented and a risk-benefit assessment, does the committee find that the Ceprate device is safe and effective for selecting CD34 positive cells for hematoporetic reconstitution after chemotherapy?" In other words, should the panel recommend to the FDA that it approve the Ceprate system?

Amy gripped my arm with both hands, digging in with her nails, but I was so nervous I barely felt a thing. I stared directly at the panel, trying with great difficulty to look relaxed, to appear confident. In Seattle, just as the chairman asked those in favor to raise their hands, the people gathered in the boardroom at CellPro lost their telephone connection to the hearing. They had absolutely no idea what was happening.

I was sitting there, but I could barely believe what I was seeing.

Every hand was raised. Every hand. I looked over the panel two times, three times. The vote was thirteen in favor and zero against. The panel had unanimously voted to recommend approval of the Ceprate system.

I cleared my throat. I wanted to thrust my hands into the air in celebration, but the hearing was still in session. On our side of the table, we all fought to suppress our excitement. I looked at Cindy, who was on the verge of tears. She had this huge smile on her face. I felt as if I were going to laugh *and* cry.

Fifteen minutes later the chairman finally announced that the business of the morning was completed, and adjourned the session. With that, the ballroom erupted. It reminded me of a courtroom scene from an old black and white movie, in which reporters raced out of the courtroom as the verdict was announced to call the news into their papers. But in this case, rather than reporters, it was stockbrokers and analysts who raced into the hall and speed-dialed their offices on cell phones with orders to buy. People in that hallway were probably exposed to more radiation than inside a microwave oven. Literally within minutes, our stock began rising. The first cell-separation system in history had been recommended for approval. We were back in business.

While the FDA would only approve use of the Ceprate system for stem cell purification, in fact we were involved in more than sixty different clinical trials. Just as Ron Berenson had once dreamed, clinicians were using his system in experimental treatments for an extraordinary variety of diseases. Prestigious institutions like the Hutch, the National Institutes of Health, and Columbia University were using it in gene-therapy protocols, inserting genes into stem cells to treat such inherited diseases as Gaucher's disease and Severe Combined Immunodeficiency Syndrome, "boy-in-the-bubble disease." At the Children's Hospital of Michigan in Detroit, doctors transplanted *in utero* purified stem cells carrying healthy genes into a four-month-old fetus previously diagnosed with SCIDS. Five months later this baby was born with a functioning immune system. These transplanted genes safely reconstituted the infant's blood and immune systems.

In May 1996, Dr. Richard Burt at Northwestern University Medical School had begun using autologous marrow transplants to treat multiple sclerosis, a disease caused when a patient's own immune system suddenly attacks the myelin sheath that covers nerve fibers. Dr. Burt believed a stem cell transplant could slow the progression of the disease. Other physicians at Northwestern as well as at the University of Wisconsin were conducting similar clinical trials to treat other autoimmune diseases, including lupus and rheumatoid arthritis.

At St. Jude's Children's Hospital in Memphis, and at the City of Hope outside Los Angeles, researchers were using the Ceprate system in experimental protocols to treat AIDS.

In Europe, researchers were using it to combat organ rejection in solid-organ transplant patients. By giving recipients stem cells from the organ donor they were trying to modify the patient's immune system enough to accept the donated organ.

It had long been believed that diabetes could be treated by transplanting donated insulin-producing pancreatic islet cells, but in reality that had not been possible because the recipient's immune system treated the donated cells as the enemy and attacked them. At the Diabetes Research Institute in Miami, researchers were hoping to transplant the donor's stem cells in addition to islet cells, with the hope of increasing the recipient's tolerance of the transplanted pancreatic cells.

In October we received permission from the FDA to begin T-cell depletion trials in childhood leukemia patients. That would allow these children to be given stem cells from family members who were only partially genetically matched. If successful, someday the T-cell depletion system might allow patients with no genetically matched donors to be safely transplanted.

It was obvious we'd given transplanters an important new tool. It seemed as though almost every day we were hearing from doctors who had found some exciting new application for the Ceprate system. One of the things that made CellPro unique was that we were thrilled to work with these doctors. If they had a promising concept they wanted to explore, we were willing to support new clinical trials. It was obvious that we had an exciting new medical technology with numerous potential applications. We'd proved it worked. But what we'd accomplished was barely the beginning.

It just didn't seem possible to me that this technology would be allowed to disappear in a business dispute.

We were caught in a trap of our own making. We had proved that there existed a potentially vast market for a cell-selection system. Our success had made Civin's patents very valuable, and Baxter expected us to pay for that success. But our profits were still in our future; there wasn't one day when we actually made money.

And until we received official FDA approval of the Ceprate stem-cell concentration system, we were going to continue to burn cash. We were in an unbelievably frustrating situation. We had panel approval, the labeling had been completed, but without final FDA approval we couldn't market the product. Financially, for a small company that had only one product, it was murderous. We were spending $1.5 million a month just waiting for what was essentially paperwork to be completed. Since the special panel had signed off in February, we'd gone through almost $12 million. We had hoped to announce the product launch at the annual ASH meeting in December, but it looked as if the FDA was moving too slowly for that to happen. Finally, against the better judgment of our regulatory people, I called someone I knew at the FDA and basically begged for help.

The FDA moved into hyperdrive. Unfortunately, hyperdrive for the FDA is still very slow. Frustratingly slow. Knowing that we might receive approval at any moment, we brought two sets of display graphics for our booth at ASH: one set to promote the system if we were approved, the other set if we still had not received approval. Monica Krieger, our Director of Regulatory Affairs, stayed in Seattle in case the FDA needed additional information.

On December 6 we were scheduled to host a symposium to report on the progress in stem cell therapy. As the symposium began, I was standing way in the back of the room, next to the door, waiting to hear from Monica. I had been confident we would receive official approval before the symposium began; in fact, we had intended to announce it there. The room was filled with hundreds of physicians, every one of them a potential customer who wanted to learn more about advances in stem-cell therapy. It was incredibly disappointing to have in that large room all the people we needed to make the product launch a success, and not be able to launch it.

I cursed the bureaucracy. Suddenly my silent beeper began vibrating. Monica Krieger was calling. I practically ran to the phone to return her call. She told me, "We got it, Rick. The approval just came in on the fax. You can go ahead and announce it." The cavalry had arrived in the nick of time. The FDA had officially approved the Ceprate SC system for purification of stem cells in bone-marrow transplants.

We were in business. Almost two and a half years after the FDA had essentially warned us there was no way in hell we were ever going to get this product approved, we had that approval. They had been the longest two and a half years of my life, but it seemed only an instant ago that Berenson had called to tell me we had been turned down by the FDA. I was so excited I wanted to shout the news out loud, but I held myself back. For the transplant community this was very significant news, and I wanted it to be announced in the best possible way.

Our next scheduled speaker was the University of Colorado's Dr. E. J. Sphall, who had been the principle investigator in our Phase I and II trials, and had participated in the Phase III trial. She'd been using the Ceprate device to treat patients for a long time. There couldn't be a more appropriate person to announce this news.

As she was walking toward the podium, I stopped her and took her aside. "E.J.," I said excitedly, "you ready for this? We just got the approval. Why don't you go ahead and make the announcement."

I stood on the side of the room as she reached the podium, feeling in some ways the pride I'd felt when my boys were born. When I'd joined CellPro, more than five years earlier, the entire company easily had fit into the corner of a small building. We'd been through so much since then. There had been many days when our survival was in question. But we had survived, we'd done it, and we'd grown to be strong. Now we had our own modern building, we had a manufacturing facility, we had 160 employees, we had established a presence in Europe—and we had an extraordinary cell-selection system that could help make possible medical wonders.

A few minutes into her talk, E.J. announced that the Ceprate system had officially received FDA approval. It was a very large room, but it was completely silent for several seconds. Then a small group of CellPro people standing in the back started cheering and clapping and hooting. It was a pretty noisy display for a symposium discussing potential therapies for debilitating and fatal diseases. But after everything we had been through for so many years, this was a sweet, sweet victory.

Until that moment the Ceprate system could be used only in experimental protocols. We had not been allowed to sell it. We were only permitted to recover the cost of the device. That was done. Now we could sell it like a can of soup. The following morning our market launch began. We were open for business.

But that was only the beginning. The next day Ollie Press mortified Nicole and Sharon, who were sitting in the audience, and electrified just about everybody else when he presented the Rick Project data publicly for the first time. Most of the people in attendance knew about our Ceprate system, but very few of them knew that we had successfully developed and utilized a B-cell depletion device. There was a lot of interest in the data. This was another important step in the development of cell therapy, a means to obtain tumor-free stem cells. If we were able to deplete tumor cells, there was every reason to believe we could also eliminate breast-cancer cells and the T-cells that complicate allogeneic transplants and just about any other undesired cell. By the time ASH was over, the phones at CellPro began to sing the beautiful music of success.

I knew it would be years before the B-cell depletion system received FDA approval. But after some further refinement in our laboratories, our next step was to get it into the hands of clinicians who would use it to treat patients. We had an extremely strong clinical development program in place. We had doctors who would take our technology and use it every day in a variety of conditions. They'd determine what worked, they'd suggest improvements, they'd tell us why it failed. This field testing was the medical version of turning a '75 Chevy into a hot rod. The results of these trials would be reported to us, and, based on this data, we would continue our development program. At some point we would meet with FDA representatives to determine an end point for approval—what beneficial results would we have to show in clinical trials before being allowed to market the depletion system?

But eventually an improved version of the system that I believed contributed to the success of my transplant would be readily available to anyone who needed it. It would help save lives.

None of that affected Judge McKelvie. After granting Baxter judgments as a matter of law on two major issues, over the next six months he entered summary judgments on six additional issues. Four days before the second trial was scheduled to begin, Coe Bloomberg called me. "Guess what?" he said. "McKelvie just eliminated our last and best defense." It was incredible, but none of us were surprised. By that point we had come to expect it.

We had believed that our strongest defense was that Civin had not enabled his patent, meaning that he hadn't taught people of "ordinary skill in the art" how to do it. In fact, even he couldn't do it again in his own lab. McKelvie dismissed the fact that Civin's own lab was unable to make another antibody by explaining,

> *Testimony at trial established that a person skilled in the art of mak-*
> *ing monoclonal antibodies must have a bachelor's degree in the appro-*
> *priate scientific field and must have made a monoclonal antibody at*
> *least once. . . . Testimony established, however, that many of the*
> *people working in the lab were undergraduates or had never made a*
> *monoclonal antibody, as evidenced by the fact that most never even*
> *achieved a working fusion as the first step in making a monoclonal*
> *antibody. Even [one of] CellPro's experts admitted that Civin may*
> *have hired workers in his lab simply for them to practice lab tech-*
> *nique, not because they had extensive training and experience in mak-*
> *ing monoclonal antibodies. . . .*

In many ways the law is like the predictions of Nostradamus. In it one can find justification for almost any claim. McKelvie excused Civin's failure to enable his own patent by admitting his lab staff wasn't "skilled in the art" of doing exactly what they were supposed to be doing. In other words, his lab staff just wasn't good enough to do it.

That last ruling eliminated the necessity of a trial to determine the facts. McKelvie had reversed every decision reached by the first jury in favor of CellPro and supplanted them with a ruling against us as a matter of law.

The second trial began March 1997. I was back at work full-time by then, but I was still fragile. Ollie had warned me, correctly, that it would take a year for me to fully regain my strength. My hair was finally starting to grow back, but I had continued wearing my wig. The Sunday before the trial started I was working out in the gym and I just didn't want to wear that thing anymore. Screw it, I decided. The wig weighed only a few ounces, but once I'd made that decision, it felt like I'd finally rid myself of a huge weight. I immediately got my hair neatly trimmed. It was amazing what that haircut did for my morale.

Late that afternoon I went to a strategy session with our lawyers. When I walked into the room, Tom Kiley looked at me once, twice, and did a triple-take. "What did you do?" he wondered.

"I got rid of the piece," I told him. "I don't need it anymore." For me, this was a moment of great triumph. I had made it back. It was ironic that it occurred at the beginning of our legal defeat.

Patty had come with me to Delaware, and the moment I walked back into the courtroom the complex array of feelings generated by the first trial came rushing back. The difference was that this time it would take a miracle for us to win. The best we could hope for was that CellPro would survive.

The scene was surreal. This was the same courtroom in which the first trial had taken place. With the exception of the jury, many of the same people were present—although the task force of lawyers that had represented the opposition in that trial had been reduced to Ware and one or two others. The last time I'd seen Ware, he'd been sitting alone at the same table, dutifully collecting his papers. I glanced at him, but if he even noticed me he didn't respond.

When McKelvie entered, I imagined he was walking with a more spritely step. He said a few words to his clerk, and then began this trial as if the first trial had never happened, as if this were to be a real finding of evidence. His demeanor was so professional that someone walking into the room for the first time could never have realized exactly what was taking place.

As far as I was concerned, this entire procedure was a sham, a kangaroo court. The only thing that was lacking was a blindfold and a last cigarette. It had all the trimmings of a real trial; it took place in a courtroom presided over by a judge in front of a jury, the plaintiffs and attorneys called witnesses—among them, to my surprise, was FACS expert Mike Loken, who testified for Baxter—but it was about as real as pro wrestling. As often as I have thought about this trial, it remains a terrible nightmare. When I state the facts as I know them, people believe that surely I must be leaving out pertinent information, or that my admitted bias prevents me from understanding the actions of the judge. But I was in that courtroom. I watched this tragedy unfold. At times I wanted to cry.

I knew it was over as I listened to Judge McKelvie read his instructions to the jury at the very beginning of the trial. Having already ruled that we had no defense against Baxter's claims, Judge McKelvie conducted the second trial only to determine damages and whether we had willfully infringed Civin's patents. The judge told the jury to find that we had willfully infringed those patents if Baxter, Becton Dickinson and Johns Hopkins could prove, by clear and convincing evidence, that we "had no reasonable basis for believing that the patent claim[s] at issue [were] invalid or not infringed or unenforceable."

He then told the jury that in prior proceedings he had ruled that "no reasonable jury" could conclude that CellPro did not infringe the patents or that the patents were not valid, while precluding us from informing this jury that in the first trial after hearing all the evidence, a jury had ruled for CellPro and found that the patents were invalid *and* not infringed.

I sat there with Coe Bloomberg and Bob Weiss, just shaking my head. My sense of outrage was overpowering, but there was nothing I could do to express it. Both Coe and Bob knew how bad this was going to be, and had tried to prepare me for it. But old beliefs die hard; I kept thinking that this couldn't possibly be happening. We went through all the motions. At night, just as we did during the first trial, we would all gather in our "war room" to prepare for the next day. This time Patty was working with us. On one occasion we worked through the night preparing notebooks for the jury, detailing precisely what CellPro had done on its own. The next morning Ware objected, and McKelvie wouldn't even let them see the books.

Patty was amazing. She ran errands, she worked with the paralegal staff, she got up early to make breakfast, and she'd sit in the courtroom the entire day. And even Patty, who finds something nice to say about everyone, could find nothing redeeming to say about this exercise. "Can they do that?" she would ask. "This is unbelievable. What does this judge have against us?"

Questions that had no answers. By the end of the trial I had nothing but disdain for the judge. Lives were at stake in this trial, and he would sit there and make flippant comments. At one point, for example, I think Coe was arguing a minor point with Don Ware and McKelvie said something to the effect that "you guys can argue about this all you want, I'm appointed for life. I'll wait for you. . . ."

The other side would laugh at remarks like this while I sat there thinking, *He doesn't get it, he just doesn't get it.* By the time Coe made his final arguments, he was a beaten man. I had never seen him like this before, either in the courtroom or in private. His voice had absolutely no strength, no energy. Several times, members of the jury had to ask him to speak louder. My belief is that his despair was not just because he had lost this case, but rather for what had been done in the name of a legal system for which he had had so much respect.

In his closing arguments, Don Ware summed up the plaintiff's case. According to this version, Tom Kiley was the mastermind behind a sophisticated fraud. Supposedly, Kiley founded CellPro to make a lot of money. The only potential roadblock to CellPro's success was the Civin patents, so Kiley had persuaded his friend Coe Bloomberg to write some bogus opinions stating that those patents were invalid. Kiley had used that opinion to raise financing for CellPro, then had eventually sold his stock and left the company to suffer the consequences.

Tom Kiley may be accused of arrogance and occasional bullheadedness. But patent fraud? Absolutely no way. Willful infringement? Absolutely no way. And even to suggest that Coe Bloomberg would risk the reputation he'd spent a lifetime building to help an already wealthy friend make a few more bucks is beyond belief. But that's what these men were accused of doing.

As I sat in that courtroom listening to this nonsense with disbelief, I felt about as helpless as I had while lying in bed waiting day after day for white cell counts. They were killing my company, and I was powerless to do anything about it. At times I wanted to stand up and scream at the injustice. The only decisions left to the jury were whether we had willfully infringed the patents, and the amount of damages we would pay. Baxter claimed $2.3 million in damages, even though we had never made a penny of profit and they did not have an approved product. On March 11, to the surprise of absolutely no one who had followed this proceeding, the jury found that CellPro had willfully infringed two Hopkins patents. They awarded the plaintiffs the entire $2.3 million in damages they had requested. Baxter, Hopkins, and BD immediately asked the court for treble damages, which would be $6.9 million, in addition to approximately $7 million for attorneys' fees.

"This decision was not surprising," reported the Wall Street securities firm UBS, which followed CellPro for its clients, "since the judge had removed virtually all of CellPro's defensive strategies in a series of summary judgments preceding this trial. . . . This adds another improbable aspect to this increasingly bizarre case."

Having gotten just about everything they had asked for, Baxter wanted more. In a post-trial motion, the plaintiffs also asked McKelvie for an injunction that effectively would take the Ceprate system off the market—as well as preventing us from continuing research and development. Don Ware wrote this injunction, and McKelvie implemented it just as it had been handed to him. McKelvie told us that if this created a hardship, we should return to court.

We returned immediately, pointing out that the injunction prevented us from doing basic quality control, which made it impossible for us to function. The injunction was modified to permit us to do quality control—but not R&D. The injunction prevented us from using our antibody in our laboratories. Effectively, our R&D program stopped.

Perhaps it was not in the purview of the court to take into consideration the results of this injunction on real people. Maybe the judge felt this was the legally proper thing to do, that there was no place in a courtroom for compassion; or perhaps he just didn't realize the effect this injunction would have on human beings. It was my feeling and only my feeling that this injunction was vindictive. It was punishment for the fact that we had upset the system by winning the first trial. That just couldn't be allowed. What possible harm could have been done to Baxter by allowing us to continue our research and development program while our appeal was in progress? Baxter insisted publicly that patient care would not be affected if their injunction was granted. "Baxter has no intention of denying any patient or physician access to technology which can help treat cancer," a spokesperson explained. "Our intent is to assure a smooth transition to a licensed technology."

Except that there was no other licensed technology. The threat that the Ceprate system would be removed from the market was very real. UBS Securities reported to its clients, "Under normal circumstances removing a potentially lifesaving product from the market would not be a viable option but, through what we believe to be a series of clever maneuvers by Baxter designed to convince the judge that their potentially competitive product, Isolex, would reach the U.S. market soon (i.e., they quickly filed a PMA for Isolex with the FDA with minimal U.S. clinical trial data—an application which we believe has a poor chance of getting approved) the waters have been so muddied that an injunction becomes a real possibility."

While Baxter eventually modified the specifics of their requested injunction several times, their proposal still required a gradual phasing out of the Ceprate system, and until that happened we were forced to pay them at least two thousand dollars per kit, with the exception of those supplied in support of FDA trials. Under those restrictions it would be virtually impossible for us to stay in business.

It had taken the jury only a few hours to reach their verdict. Sitting in the courtroom as the jury returned with its verdict, I was probably in a state of shock. It was no surprise to anyone that the jury found that we had willfully infringed the patent. Considering the instructions the jury was given, and the case we were allowed to present, the jury reached the only verdict possible. If I were sitting on that jury, I probably would have reached that same verdict. But still, as I heard the clerk read the verdict, I felt as if someone had stuck a dagger in my heart.

After the verdict, Coe Bloomberg, Bob Weiss, and I walked slowly back to the hotel together. No one said a word. We were all lost in storms of outrage. Suddenly, Weiss just threw his heavy attaché case on the sidewalk and angrily kicked it as hard as he could.

It was about the most appropriate comment possible. Bloomberg and Weiss were experienced trial lawyers. They understood the inner workings of the system. But neither of them had ever experienced anything quite like this, and they just didn't know how to respond to it. This was the only way Weiss could show his disdain for the proceedings.

He took a deep breath to collect himself. Neither Coe nor I said a word. Then Bob picked up his briefcase and we continued toward the hotel.

That was a terrible night. Jamie and his friend, Becky, had driven down from Worcester for the last few days of the trial and were with us. Later, at the hotel, we met someone who worked in the courthouse, a person with whom we'd become friendly during the first trial. He was very upset. He had been over to the Dupont Hotel, he explained, where the Baxter people were celebrating their victory. "They're over there drinking champagne," he said, then he shook his head and added, "They have absolutely nothing to celebrate. McKelvie handed them that case on a silver platter. He was their best friend in there."

In the past the courts had always accepted the fact that a company did not willfully infringe a patent if it hired excellent attorneys who did good diligence and wrote valid opinions. We had taken precisely the same actions that countless other companies had taken in the past. Lyon & Lyon was one of the best patent firms in the nation, yet the jury had found that we had willfully infringed the Civin patents. This was setting a potentially very dangerous precedent.

At times I found myself wondering about the people on the other side of this case. Long ago, long before I had started working at CellPro and become involved in this case, I had understood that no matter how certain I was that what I believed was right, there were people who believed precisely the opposite, and believed it as strongly as I did. That had been hard for me to accept. And in this case, almost impossible. Did Baxter really believe that we were terrible people interested only in profits? Did they honestly believe that no patient would suffer if the Ceprate system was taken off the market? Did McKelvie truly believe that?

But if they didn't believe those things, how could they try so hard to destroy us?

So I had to accept the fact that they truly believed they were right; that McKelvie believed that he was following the law as he read it in order to right what he considered a bad verdict by the first jury; that Baxter believed we were terrible people trying to steal their technology, and no patients would suffer because of this dispute. I had to accept it because if that wasn't true, I didn't know how they could live with themselves.

We left Wilmington the morning after the verdict. Coe and Bob Weiss were already working on the appeal. I had been assured that there were numerous grounds for an appeal. But having seen the circuit court system in action, I had little faith in the court of appeals—particularly after having been told about the rumors that McKelvie was considered the strongest candidate for the next open seat on that bench.

The survival of the company was very much in doubt. We certainly couldn't depend on the appeals court. We had several potentially valuable products in development, but we had only one approved product, the Ceprate system for cell selection. And for business purposes, this verdict took that away from us. It was almost impossible for me to accept that this dispute had reached this point. I, more than anyone in the world, knew the value of our technology. I was literally living proof that it worked. When we stated publicly that we might be forced out of business as a result of McKelvie's decision, many people believed it to be nothing more than a threat. But it wasn't a threat. It was cold reality.

McKelvie had already decided that the Ceprate system infringed Civin's patents. The only determination left to be made was how he intended to assess damages. A jury would make that decision, but the fact that the jury would be sitting in Judge McKelvie's courtroom did not inspire great confidence. If we were prohibited from using the system, which was unlikely because McKelvie wouldn't keep this technology from dying patients, or if Baxter was awarded large monetary damages as well as a significant share of the proceeds of every sale, which was more likely, we wouldn't be able to stay in business. As a public company with little prospect of ever earning a fair profit, our stock would have almost no value. We certainly wouldn't be able to attract the new investors necessary to keep the company growing.

At times we discussed substituting a different antibody in the system, an antibody that Baxter could not claim infringed their patent. Technically, it may have been possible. Legally, it would have made Baxter's case moot. Financially, it couldn't be done. It had taken CellPro more than five years and more than $100 million to get our cell-separation system approved. If we substituted a different antibody, we would have to repeat the entire approval process, starting once again with Phase I clinical trials. Even if we had wanted to reinvent the company, we couldn't have done it. The once-sizzling stock-market interest in biotech companies was long done, a victim of too many product failures, the extensive length of time it took to get a product approved and to market, and excitement over Internet companies. We had been founded to exploit a stem-cell-processing technology and had done so successfully, and we were going to survive or perish as a stem-cell company. We had no options.

When it had become obvious that we had absolutely no chance of winning in Judge McKelvie's courtroom, we had begun to explore other options. We followed many different paths. Some of them reached dead ends; others can't be revealed because they are covered by confidentiality agreements that may or may not remain in force. But we engaged in discussions with several other companies about many different solutions. At that point we were willing to do almost anything to stay in business, but potential partners were frightened away because we still didn't have a decision from McKelvie.

We had finally run out of options. Our survival was at the mercy of the court, and seemingly that court did not want us to survive. And then Coe Bloomberg found one final path, a path that had never before been followed. Civin's invention, as well as the work originally done at the Hutch by Ron Berenson, had been licensed to private industry under the terms of the Bayh-Dole Act of 1980. Before passage of that act, the government annually invested as much as $30 billion in university research and got very little return. By 1978 the government owned 28,000 patents that were pretty much just sitting on the shelf. The bill permitted universities to patent discoveries made with tax dollars and sell exclusive licenses to private industry. Few congressional bills have had more impact on real life. In addition to practically fostering the biotech revolution, through the years Bayh-Dole has generated billions of dollars in royalty payments to universities. It transformed the once-esoteric world of the university laboratory into an educational profit center. Before the bill was passed, about 150 patents were issued each year to universities; following passage of Bayh-Dole, universities have patented almost two thousand inventions and discoveries a year.

Few people outside the biotech industry are aware of the existence of Bayh-Dole. While it has fostered life-enhancing technology, it has also been a gold mine for industry. Billions of taxpayer dollars are being used to support university research that is then being sold to private industry, where it is developed further and sold back to the taxpayers who paid for it in the first place. Estimates are that as much as 25 percent of the research and development costs associated with bringing a product to market currently are being paid by the government. In some ways, Bayh-Dole transformed university laboratories into R&D labs for private industry.

While the bill was being debated by Congress in 1980, there was legitimate concern that big corporations would take advantage of it to gain a better competitive position. Several senators opposed to it described it as corporate welfare for the pharmaceutical industry. They were afraid that large companies would license promising technologies, then gouge consumers by making them pay exorbitant prices for products developed with government funding, or these big corporations would buy patented technology simply to prevent competitors from obtaining it, then let promising inventions linger without being developed. They were concerned, for example, that Bayh-Dole would let a big oil company buy the rights to an invention that enabled cars to get one hundred miles from a gallon of gasoline, then refuse to market it.

To prevent that from happening, legislators added several limiting provisions. As a way of using Bayh-Dole to encourage the growth of small businesses—like CellPro—the bill provided that smaller companies were to receive preferential treatment "when licensing a subject invention if the contractor determines that the small business firm has a plan or proposal for marketing the invention which, if executed, is equally likely to bring the invention to practical application . . . provided that the contractor is also satisfied that the small business firm has the capability and resources to carry out its plan or proposal."

And to ensure that major corporations couldn't buy and bury promising technologies, Congress added what became known as the "march-in provision," "to ensure that the Government obtains sufficient rights in federally supported inventions to . . . protect the public against non-use or unreasonable use of inventions."

In simple terms, the government retained the rights to make sure that any invention for which the public had paid remained available to the public. The government could "march in" and grant licenses to technology if a situation existed that had a negative impact on public health or the government deemed it to be in the best interests of the American public. Bayh-Dole stated that the government could "march in" and force a patent holder to sell a license to a responsible applicant on reasonable terms if it was determined that:

> (a) action is necessary because the contractor or assignee has not taken, or is not expected to take within a reasonable time, effective steps to achieve practical application of the subject invention in such field of use; [or]
> (b) action is necessary to alleviate health or safety needs which are not reasonably satisfied by the contractor, assignee, or their licensees.

Johns Hopkins applied for a patent for Civin's invention in 1984, and licensed the technology to Becton Dickinson in 1985 for all applications. BD eventually decided to keep the diagnostic rights and sell the therapeutic rights. They sold those exclusive rights to Baxter for, I believe, a $1.5-million payment and a 5.5-percent royalty—a significantly higher rate than they were paying. They had added very little additional value to the patents. It was obvious that BD had not taken steps within a reasonable amount of time to turn Civin's discovery into a product. In fact, all they had done was sell it. Baxter then turned around and offered *nonexclusive* licenses to anyone working in the field for an effective 8 percent royalty plus an up-front fee of $750,000. They had added no value at all. They were simply reselling, at an even higher price, the license they had purchased from BD. Eventually, after years of negotiations with us, they increased the price to "double-digit millions" plus a royalty. This is known as "layering." And that was certainly not the intent of Bayh-Dole.

While Baxter had spent six years attempting to develop its Isolex cell-separation system, it was nowhere near gaining FDA approval. Just as important, Baxter had announced its intention to sell the division that was developing and testing its stem-cell separation technology. Clearly it didn't even intend to be in the stem-cell business in the future. But Baxter had threatened to request in court that the Ceprate system be removed from the market if, as appeared likely, it won the lawsuit. That alone seemed to me proof that there existed a health or safety need that was not being satisfied by the Isolex system.

Like most people in the biotech industry, I had been aware of Bayh-Dole, although I didn't know all of its many provisions. I certainly didn't know anything about this "march-in." Coe Bloomberg had first brought it to my attention after McKelvie threw out the verdict from the first trial. But only after the judge began dismantling all our defenses did we begin to seriously consider it. As I read the bill, it seemed as if it had been written expressly for this situation. We had the only FDA-approved stem-cell system in the world. The Ceprate system had already been used to treat about five thousand patients. Baxter had not even begun Phase III clinical trials. The other companies to whom Baxter had sublicensed the Hopkins patents, Applied Immune Sciences and SyStemix, had not developed viable products. By threatening to have the Ceprate system removed from the market, Baxter was certainly proposing to deprive the American public—the people who had funded both Civin's work and Berenson's work—of cell-separation technology.

When it became obvious we weren't going to win, we decided to pursue a second strategy. We would take our case directly to the people who would be most affected if we were forced out of business, the American public. It seemed to me that most people would be as outraged as I was if they were made aware of what was happening. How could the government turn us down? We were a responsible small company with proven technology, and we were willing to pay even more than the "reasonable terms" described in the bill.

Bayh-Dole seemed like a possible way out of this mess. If we were successful, the government could grant us a license to Civin's patents or force Baxter to sell us a license at a fair price. In a sense the government would act as a mediator to force a settlement in this dispute. We would get the license, Baxter would get a fair return on its investment and the public would get the benefits of our technology. Who could possibly be against that? All we had to do was convince the government to "march in."

But as we began investigating, we discovered an incredible fact: since the passage of Bayh-Dole, the march-in provision had *never* been invoked. In fact, Lyon & Lyon had to do quite a bit of research just to figure out how to proceed. In this case, enforcement of Bayh-Dole was the responsibility of the Department of Health and Human Services. The decision would be made by the Cabinet secretary who ran HHS, Donna Shalala.

I really had very little understanding of how difficult this task would be. In addition to being a poorly defined and complicated technical process, it was just dripping with political implications. We were challenging one of Washington's most important sources of power, the big corporations who contributed freely to political campaigns. Trying to force the bureaucracy to move against its will was like pulling a reluctant elephant on a dog leash. To shepherd us through the process, we hired a Washington law firm particularly familiar with Bayh-Dole, Wilmer, Cutler and Pickering. We also enlisted the help of the man who had originally cosponsored the bill, former senator Birch Bayh.

On March 3, 1997, Bayh and our attorney, former White House counsel Lloyd Cutler, petitioned Secretary Shalala to exercise the march-in provision "to require that a license be issued to the extent necessary to ensure that (the Ceprate system) remains on the market."

Our eighteen-page petition was a very powerful statement: "The exclusive licensee's [Baxter] own efforts to develop a product for such use [breast cancer and related diseases] has not resulted in an FDA approved product and may never do so. Nevertheless, the licensee has refused to license our client's product on reasonable terms and is taking the position that its approved product should be subject to an injunction preventing its use." Later in this petition, Bayh and Cutler emphasized that "Baxter's counsel stated at a pre-trial hearing last week that Baxter intends to seek a permanent injunction to have CellPro's products removed from the market if it prevails in the retrial. . . . In doing so, Baxter threatens the welfare and very lives of many individuals who need bone marrow transplants and whose suffering could be lessened and whose lives could be saved with these products. The Secretary has the authority under the applicable law and regulations to avoid this result. . . ."

The petition recited the entire history of this dispute, including all attempts to negotiate a settlement. We noted that in the first trial Baxter's expert witness had claimed that a reasonable royalty for us to pay them would be $15 million—plus 16 percent on sales of all our products. That was a figure that would have made it almost impossible for us to stay in business. We added that Baxter had modified that position in preparation for the second trial, but when we'd proposed a settlement, they told us they weren't interested because they expected to win that trial—and when they did, we would have to take our products off the market.

Secretary Shalala could not have been less interested in getting involved. She handed the problem to Harold Varmus, the director of the National Institutes of Health, reasoning that NIH had provided the original funding for the technology.

We certainly couldn't match our adversaries' political clout—both BD and Baxter were multinational giants, and Johns Hopkins receives more government support than any other American university—but we could rally public support. So, in addition to law firms, we hired the public-relations agency Burson-Marsteller, experts in crisis management, to help us lobby for "march-in." We'd won the battle of the molecules; this was the beginning of a war of words. It was to be fought in front of millions of people on television and in the pages of the nation's most important magazines and newspapers.

Ollie had warned me to go slowly, but at CellPro the only speed was pedal-to-the-metal. It seemed as if we were always racing at full speed. Six months after leaving the hospital, I was back at work full-time plus. I just didn't have the time to be sick anymore. We had filed our petition asking HHS to begin the march-in procedure just as the second trial began. That timing aptly describes the way we felt about our chances in this trial.

While we waited for Judge McKelvie to issue his final ruling on damages, we concentrated our efforts on the march-in petition and public relations. It very rapidly became obvious that the government did not want to get involved with this problem. Several people told me that this was the most politically sensitive issue the NIH had ever faced. At one point we even contemplated filing for bankruptcy, based on the long-term effects of McKelvie's decision, as a means of forcing HHS to march in.

We tried very hard to schedule a meeting with HHS Secretary Donna Shalala, but she just did not want to meet with us. I had everyone I knew who had contact with Secretary Shalala try to arrange a meeting, but it never happened. Even after receiving a letter written to her by Birch Bayh, in which he wrote, "For those of us who have lost a loved one to cancer, the CellPro technology is a miracle. Hopefully, we can work out the legal issues without the unnecessary loss of lives," she still was unable to find a few minutes to speak with us. The only officials at HHS willing to speak with us were attorneys in the licensing department at NIH. Nice people, but they had absolutely no power to help us.

Another aspect of my education had begun.

At the urging of Burson-Marsteller, we decided to try to convince legislators to pressure NIH to approve our petition. Birch Bayh, Dr. Richard Burt, and I spent a day walking the corridors of Capitol Hill, trying to lobby for support. We met with just about everyone who was willing to speak with us. Burson-Marsteller suggested we bring with us someone who was actually using the system, and Richard Burt had agreed instantly when we'd invited him. Richard Burt is a quiet, understated, extremely dedicated doctor. We'd asked him because, before establishing himself at Northwestern as one of America's leading transplanters, he'd worked at the National Institutes of Health. The theory that autoimmune diseases could be treated successfully with stem-cell transplants was not new, but Richard Burt had proved that it could be done. Using the Ceprate system to treat patients with advanced multiple sclerosis, he had gotten extremely promising results. He believed he was on the verge of a significant medical advance—which would be lost or delayed if we went out of business.

I remember one meeting in particular with a member of the staff of Carol Mosley-Braun, at that time a senator from Illinois. I'm certain that the only reason this aide agreed to meet with us was that Dr. Burt was on staff at Northwestern, but since Baxter's corporate headquarters were in Illinois, I had very little hope that Mosley-Braun would support us. If there was any doubt that the senator would support a billion-dollar company with headquarters in her state, it disappeared immediately. "Patent infringement is a terrible thing," this aide explained, "and we believe that this is really a corporate fight. There's just no reason we should be involved."

In response, Richard Burt very softly began talking about his work. "Look, I'm just a transplant physician," he said, "and I don't know anything about corporate issues. All I know is that I've been treating patients with this technology and I've gotten some pretty remarkable results.

"For the first time with this disease we're really helping people. We're making a difference. It's vitally important that I'm able to continue using this system in my studies. If I have to stop or switch to a completely different system . . ."

I had expected to be politely thrown out of the office. But as Dr. Burt continued, I could actually see the aide beginning to hear him. Meeting us had been a political courtesy, but this story had caused him to listen. Unexpectedly, a political decision had taken on a human dimension. Richard Burt's passion flooded the room. By the time he'd finished, the aide had agreed to look further into our request. The transformation had been amazing, although I knew the political reality made it almost impossible for Mosley-Braun to support us.

I saw that repeated several times during those meetings.

We had substantial success. While we expected to receive the support of legislators from the Northwest because we were their home team, we were gratified that several senators actively got involved because they thought it was the right thing to do. Eventually ten senators and twenty-five members of the House co-signed a letter to Secretary Shalala stating, "We are writing to you on behalf of breast cancer victims, leukemia patients, multiple myeloma patients and many others across the states and the nation who suffer from cancer. It is our understanding that a pivotal new tool in the fight against these diseases . . . may soon face removal from the marketplace. . . .

"It is our understanding that an alternative product is a long way off from receiving FDA approval. If this is indeed the case, it would be a horrible injustice to American cancer victims and their families if their access to the Ceprate system was denied for a non-medical reason. . . ."

New York's Senator Alfonse D'Amato held a press conference and issued a release stating, "We can't sacrifice the health of women to the profits of companies. . . . Cancer patients should not be forced to put their treatments on hold while the legal process runs its course."

In addition to politicians, more than thirty cancer specialists who had used the Ceprate system submitted declarations on our behalf. These doctors weren't CellPro employees, and they had nothing to gain by supporting us—except perhaps the continued use of our technology. Dr. Kent Holland from Emory University wrote, "Its removal from the U.S. market would effectively remove potentially life-saving therapy from patients who cannot afford to travel abroad for treatment."

Dr. Andrew Yeager, also from Emory, wrote, "If for any reason the CellPro T-cell depletion device were to become unavailable this study would need to be shut down. If that were to happen, children would die."

Dr. Claudio Anasetti from the Hutch explained, "The ten mismatched-donor transplant patients we treated could not have been treated without the use of the Ceprate column. If not treated these patients would have died. . . ."

To counter these statements, Baxter submitted to the court a declaration from Dr. Scott Rowley, whose staff had done the actual processing of my blood, in the Rick Project. When I read Rowley's statement I was absolutely stunned. His statement claimed that Baxter's Isolex system was superior to CellPro's technology.

Baxter used this declaration in court and in its own p.r. campaign.

I called Scott immediately. And during that phone call he revealed to me, for the first time, that he was a paid consultant for Baxter. I felt completely betrayed. We never spoke again.

That wasn't the only surprise that Baxter had in store for me.

On April 27 the president of the American Cancer Society wrote to Secretary Shalala, "We believe that an injunction in the absence of a comparable FDA-approved alternative could have a devastating impact on thousands of patients whose therapy could depend on [the CellPro] system." Interestingly, two months later the president wrote again to Shalala, basically rescinding that request because he had been assured that "ongoing negotiations between CellPro and Baxter" had resulted in agreements that ensured the Ceprate system would remain available until Baxter's Isolex system received final approval from the FDA.

The problem with that, unfortunately, was that it wasn't true. During a subsequent telephone conversation, he revealed that this information had come from Baxter. Even though none of this was true, he refused to write again to Shalala. I wrote to him, "There have been *no* ongoing negotiations between CellPro and Baxter, *and no such agreement has been reached.*"

We fought to stay in business. Larry Culver and I spent countless hours planning strategy or responding to Baxter's latest legal salvo. Every day was a barrage of phone calls, faxes, and e-mails as we tried to find a way out of our desperate situation. But we might just as well have been trying to stop the tide from coming in. When I was growing up in the 1950s, I loved low-budget monster movies. At some point in many of these pictures, someone would rush into a laboratory and warn the hero, "It's reached the edge of the city."

Baxter had reached the edge of the city.

At times I was so completely absorbed in the fight that I forgot completely that only months earlier I had been dying. Clinically, even though there was no sign of tumor cells in my system, my disease was in remission. I had done everything possible to put my disease firmly in my past. I was busy fighting a new fight now. Which brings to mind the story of a man out walking his elephant. When asked where he'd gotten his elephant, he replied, "What elephant?"

What disease? At about the same time we initiated the march-in campaign, a reporter from the *Seattle Post-Intelligencer* named Carol Smith learned about the success of the Rick Project. She recognized a good story with a strong local angle. I was very happy to cooperate. I thought a little publicity in the community might be of some help. In late April 1997, her story appeared. It was the first drop of publicity in what was to become a great flood. "When Rick Murdock first noticed the lump on his neck," the story began, "he never imagined his own company's researchers would rally and perhaps save his life."

Her story quickly attracted the attention of the national media. The phones began ringing and didn't stop for several months. We had no grand plan to use the Rick Project story to gain attention and sympathy. I wish I could claim we were that smart, but it really was a case of "What elephant?" Once reporters began calling, though, we realized immediately that this was an extraordinary opportunity to rally support for our march-in campaign. To every person who called, I began telling the entire story of my survival as well CellPro's fight for survival.

On May 1, Bill Richards of the *Wall Street Journal* made it a national story. Richards had been investigating the story for many weeks. "Rick Murdock is betting his life on an unproved product made by his biotech company. It's possible that neither the man nor the company will survive. . . . 'People dread turning fifty and getting old,' Mr. Murdock said. 'I don't. I'm going to be fifty and all I can think of is, I made it.'"

The story had all the elements the media savors, a great human-interest story that also affected the lives of every reader or viewer. It was even better than a TV movie: A CEO's company saved his life with an experimental process—and the survival of that company, with its potentially life-saving process, is threatened in a courtroom battle over money.

Suddenly I was the subject of a tidal wave of publicity. *Time* did a long piece in which I called the loss of the Ceprate system "a crime against humanity." *Newsweek*'s story featured a young woman incapacitated by multiple sclerosis whose life may have been saved by a transplant using the Ceprate system. *People, Science,* the Associated Press, and newspapers around the country carried major stories. Tom Brokaw told my story on the *NBC Nightly News*. On *Primetime Live*, Sam Donaldson reported that "the extraordinary efforts to save one patient may eventually help thousands of people every year." CBS and Fox News covered the story. I appeared on *Good Morning America* and *Inside Edition* as well as numerous local programs. The *Today* show flew my whole family to New York, for an experience none of us would ever forget. Especially the boys, who had the opportunity to meet the band Steppenwolf.

Even if we had wanted to, we could never have afforded a publicity campaign of this magnitude. We received millions of dollars' worth of publicity. I'm sure Baxter was stunned by it. In their darkest moments, they could not have imagined this happening. In pretrial motions, Ware stated that the plaintiffs intended to ask for an injunction to take us off the market immediately, and it is my firm belief that had we not shone the spotlight on them, McKelvie would have granted that request. CellPro would have quietly gone away, and only the transplanters would have known the difference. Once this plan was publicly exposed, Ware asked the judge to stay the injunction until the FDA had approved another cell-separation product.

Faced with this tidal wave of negative publicity, Baxter, Hopkins, and BD hired their own lobbyists and public-relations agency and began fighting back. Responding to both our march-in petition and our claim that their injunction would put us out of business, Hopkins announced that steps had been taken to ensure that patients would have continued access to stem-cell purification, and that we were simply using the threat of removal to scare cancer patients into action. Basically, they were accusing us of medical extortion. A spokesman for Hopkins told reporters, "CellPro's willingness to use scare tactics to enlist the support of cancer patients and their families in its investor-driven business strategy is reckless and irresponsible."

Dr. Civin said the same thing: "[CellPro is] scaring a lot of people. We're getting calls from frightened patients and their families. This looks to me like . . . a calculated business strategy." Later he added, "As a physician who treats children with cancer every day, I am acutely aware of the value of stem-cell selection technology. I would not be involved in a dispute that might compromise the care of patients."

We weren't trying to scare people; we were telling them truth as we believed it to be. The fact that the truth was frightening did not make it any less the truth. But it was obvious that our real threat to Baxter and to the universities was financial. As had been anticipated years earlier, Bayh-Dole had indeed turned out to be a financial gold mine for universities. Licensing technology had become a multibillion-dollar business, and the threat that the government might actually exercise its rights to make sure the people got what they were paying for apparently scared many universities. In his response to our petition to Secretary Shalala, Don Ware wrote that a victory for CellPro "would send shock waves through university technology-licensing offices across the country."

A spokesperson for Hopkins pointed out, "It really goes to the heart of whether companies are going to be willing to invest a lot of money in innovations that come from universities. If we get our legs kicked out from under us," and can't guarantee that an exclusive license will remain exclusive, "I'm sure the answer will be no."

And Dr. Civin added, "If the government were to intervene in this case, it would set a dangerous precedent for the research community. If nonprofit institutions such as Hopkins cannot offer private licensees patent protection, the huge financial investments necessary to bring medical inventions to market may dry up. Patients will ultimately lose out." Essentially, their argument was that CellPro's greed could end up costing universities a lot of money.

Obviously our opponents were terribly frustrated by my willingness to discuss the success of the Rick Project. The situation got so absurd that for a while I thought perhaps they were accusing me of getting a terminal disease to try to help the company. Don Ware claimed I was milking my illness for sympathy. "They've exploited that to the maximum," he said.

Well, excuse me for living.

This really became a corporate battle to the death. For Christmas 1996 we wanted to create a Christmas card that would relate the story of a patient who had benefited from our technology. At Emory University, Dr. Andrew Yeager and Dr. Kent Holland were treating young patients with childhood leukemia. These were children who were dying. They had failed traditional therapy, leaving a stem-cell transplant as the only possible way of saving their lives. In several cases these children had no compatible family members, and a search for a matching donor had failed. Previously what had happened in this situation was that the families kept these children as comfortable as possible until they died.

The only potential solution was a stem-cell transplant from one of their parents, a haplo-identical transplant, and in the past they hadn't worked very well. Parents are haplo-identical, or only half-matched, with their children, meaning their kids have half of each parent's genes. The result of a transplant is often a fatal case of graft-versus-host disease. At Emory, Yeager and Holland were using the Ceprate system to deplete enough mature T-cells—the soldiers of the immune system—from the donated stem cells to moderate the severity of GvHD. It was an extremely sensitive process: they had to retain sufficient T-cells to promote a stable engraftment, but not enough to cause severe GvHD. No one knew what that threshold number was, so they were experimenting.

It was obvious that a T-cell depletion device could save a lot of lives. As our society becomes increasingly intermingled, it will become harder and harder to find perfectly matched stem-cell donors. When baseball Hall-of-Famer Rod Carew's daughter developed leukemia, for example, it was impossible to find a suitable donor for her, because her parents came from very different ethnic backgrounds. If our T-cell depletion system had been available, it's possible she might have survived.

The fifth person in the world to receive a haplo-identical transplant using the Ceprate system with a T-cell depletion column was a five-year-old boy. A stem-cell donation from his father was purified and a great number of T-cells were removed. The transplant worked, and the boy's terminal leukemia went into remission.

It was thrilling success story. With the permission of the boy's parents, we told it on our Christmas card, and illustrated the card with the child's own drawing of a smiling snowman. We sent the card to all of our customers, investors, and supporters, as well as to others with whom we had pleasant relations. I was so proud of the card that I sent one to our contact at the FDA, the man who had done so much to help us move forward.

That turned out to be a mistake. The Ceprate system had been approved by the FDA for one use and one use only, to concentrate stem cells from bone marrow. While most clinicians and transplanters were using it for many other nonapproved purposes, it was against FDA regulations for us to advertise or promote these "off-label" applications. Even though our T-cell depletion device was being used in approved clinical trials, for example, it was still off-label.

In January 1997, Monica Krieger, our Director of Regulatory Compliance, received a letter from the FDA informing us that we were promoting a nonapproved product and directing us to stop. This was the FDA's mildest form of rebuke. We were not required to make this letter public, and only a few people at CellPro even knew about it. But during oral arguments concerning the plaintiff's request for an injunction, in support of their argument that there was little regulatory difference between the Ceprate system and the Isolex system, they handed Judge McKelvie a copy of this letter. They claimed that the FDA had found it was "not appropriate" for the Ceprate system to be used in the treatment of this child.

By this time very few things could surprise me, but this just stunned me. The copy of the letter given to Judge McKelvie had a notation on it written by Dr. Krieger. I couldn't even begin to guess how Baxter had obtained it. This was an internal document that had been improperly divulged. One thing for certain, it wasn't Monica Krieger who had leaked it. It appeared that someone was spying for the plaintiffs, although we never figured out how they got it.

In our response we pointed out to the judge that the FDA was citing us for improper promotion and not, as claimed by the plaintiffs, improper use. And we added that in the previous five years we had never received a single FDA warning letter, while "the Plaintiffs and their related companies had received numerous such letters, including a titled 'warning letter' finding Baxter's Bone Marrow Collection kit to be 'misbranded' and threatening regulatory sanctions including seizure and/or injunction."

The fact that Baxter had gotten a copy of this letter shook me up. Both Larry Culver and I speculated at length on how deeply the plaintiffs had burrowed into CellPro.

Our march-in petition had created a complex problem for the NIH. There was really no way for them to be able to determine with certainty what damage would be done if we went out of business. The plaintiffs contended that no patient would be left without access to stem-cell purification technology. We believed that we were fighting for the present as well as the future. We were the first company in the world whose only business was to develop stem-cell transplant technologies. It wasn't simply what we had already done that would be lost if we were forced out of business, it was what we were going to do in the future. But how do you prove that to a bureaucracy rooted so deeply in regulations?

The NIH recognized that the best possible result would be a negotiated settlement, and to its credit, tried to engineer it. But by this time Baxter had won a complete victory in McKelvie's courtroom and was not about to lose it in negotiations. They wanted us to disappear. They were no longer interested in negotiating. The NIH asked both CellPro and Baxter to put forward offers that might lead to an equitable settlement. We complied immediately. Initially, Baxter failed to respond to our offer, and, when pressed by our attorneys, quickly dismissed it. They never presented their own offer.

The only other way for the government to make this problem go away was for Baxter's Isolex system to receive FDA approval. We weren't concerned about that. It had taken us five years and a lengthy and complicated randomized clinical trial to gain approval for the Ceprate system. Baxter had not even begun Phase III randomized trials, so we knew that realistically they were at least three years, if not five years, away from approval. The fact that there was no alternative to the Ceprate system was one of our primary weapons in pushing for the march-in.

And then, out of the blue, blue sky, the FDA scheduled an advisory panel meeting for the Isolex system. It had taken us twenty-six months after filing our pre-market approval application to get an FDA hearing; the Isolex hearing was scheduled only five months after Baxter had filed. The FDA had never before operated like that. Either this was one of the most incredible coincidences in history, or the FDA had decided that the solution to our march-in request was to get the Isolex system approved. That would make the march-in moot. I had even been told, although again I have absolutely no evidence of this, that the FDA had approached Baxter and asked them to submit whatever Phase I and II data they had to put in front of the panel.

Baxter actually had very little data to present. Our Phase III trial consisted of one hundred patients chosen at random; first come, first served. Baxter had twenty-two patients from nonrandomized Phase I and II trials conducted around the world. They had no control group, no consistent protocol.

The FDA position was that our tests had demonstrated the safety of stem-cell transplants and that a stem cell is a stem cell is a stem cell, and therefore the Isolex device should be approved. But even the limited data Baxter submitted was not very good, and that wasn't just my opinion.

With a potential market of about $100 million a year at stake, Wall Street was following this battle with great interest. The financial institutions were reporting facts, then interpreting them for their clients. Like doctors, these companies weren't picking sides, they were advising their clients on investment strategy. I admit to being biased; they had to be more pragmatic. So as the investment bank Hambrecht & Quist, which followed CellPro, reported about Baxter's data, "To state that this is a weak PMA (pre-market approval) filing would be a significant understatement. An extensive litany of shortcomings in the PMA filing make it a mystery as to why the PMA was accepted for review, and not rejected pending further data submission as has always been the FDA's practice."

Among the problems with this filing, according to Hambrecht & Quist analyst Richard van den Broek was that according to Baxter's own data, the Isolex system not only didn't do any good, but apparently patients who received stem cells purified by this system took longer to engraft. A six-month follow-up revealed that of four patients who died after transplant, three of them had been given Isolex-purified cells. At the end of one year, eleven patients in the study had died—seven of them having used the Isolex system.

The entire study was flawed. Substantial data was missing, and there was evidence of miscommunication between Baxter and the clinical sites, as well as several minor protocol violations.

My belief in this case was that the FDA was not in favor of approving the Isolex system. I think the pressure came down from the Department of Health and Human Services to take them off the hook by approving it. Every panel-watcher knew this hearing was a farce.

Sometimes it seems as though the whole world is the bottom of a shoe, and you're an ant. After all the battles we had waged, it was difficult to summon the energy to fight another one. I was fortunate that at the end of the day I had the ability to wind down, but Larry Culver, for example, did not. The stress took an enormous toll on him. Both of us, in fact many people at CellPro, had opportunities to bail out. CellPro had established a strong reputation in the biotech world, and people who worked there were considered top-quality. Every time we announced bad news, the headhunters descended on us. I had received several serious offers for CEO positions that paid more money and involved considerably fewer headaches.

I couldn't leave. Larry couldn't leave. Nobody could. After everything we'd been through, CellPro had become more than the place we worked. We had become a family. We'd been through the death of Nicole's baby, the Rick Project, the two trials of the Baxter lawsuit, the joy and then the collapse of the Corange deal, the initial FDA turndown and eventual approval. We'd saved lives and seen our stock rise as high as thirty-five dollars a share.

The primary accusation against us was that we had been greedy. But if money had been our primary motivation, many of us, including me, would have cashed in at opportune times. And those people making that charge, of course, had strong financial reasons for doing so. All of the people who wanted us gone would be profiting from the Civin patents.

Baxter didn't even want to be in the immunotherapy business. We had been hearing rumors for quite some time that they wanted to continue distributing products, but wanted out of R&D—the really expensive part. Eventually they did a joint venture with a company named VimRx to form a new company called Nexell. VimRx owned 80 percent of Nexell, and Baxter owned 20 percent, but by virtue of this investment Baxter also got 40 percent of VimRx. The way I understood the deal, Nexell would do all the research and development, which meant doing all the research, development, and clinical trials, while Baxter would be paid to manufacture the product, then distribute it for, I believe, a 35-percent distribution margin. Nexell would take whatever was left. At CellPro, almost 80 percent of our budget was devoted to R&D.

It was a great deal for Baxter. They were ridding themselves of the most expensive part of the operation while keeping a lucrative portion, they were getting tax benefits, and if McKelvie's decision wasn't overturned, they would also receive millions of dollars from CellPro. What I couldn't figure out was how Nexell could be profitable.

All of that made it somewhat difficult to accept the accusations that we were being greedy.

One more time we waded into battle. We filed what is known as a citizen's petition with the FDA, reminding the agency that by statute it is required to treat like parties the same way. We had been required to do a Phase III randomized trial—in fact by then we had completed *two* Phase III trials—so the Isolex system could not be approved unless they adhered to the same rules. We wanted the FDA to know we were looking over its shoulder.

We were caught in a box. Our march-in petition had put pressure on the FDA to approve Baxter's product, yet McKelvie's injunction allowed us to stay on the market only until there was another approved product. Our hope for survival was that the FDA would fail to approve the Isolex system before either our march-in petition was approved or McKelvie filed his final judgment and the court of appeals overturned the verdict.

When we filed our petition, I met with Jay Siegel and members of his staff at the FDA to explain our position. If we somehow managed to stay in business, we would be at the mercy of the FDA. I did not want this petition to poison any future relationship. "We think you've dealt with us very fairly and consistently," I told him, "but what isn't consistent is the way you're dealing with Baxter. All we ask is that you hold them to the same standard that we were held to. Because if they get approval, we die."

Siegel was very polite and completely noncommittal. The only thing that, by law, he was permitted to tell us was that Baxter's product was in the approval process, and thank you for coming.

We waited. We waited for McKelvie to issue his final ruling. We waited for the NIH to act on our march-in petition. We waited for the FDA hearings on approval of the Isolex. And while we waited, we struggled forward under the weight of the injunction with our next generation of products, particularly the B-cell and T-cell depletion devices.

I continued to speak out publicly to anyone who would listen. We were contributing too much to be executed. It was very difficult for me to fathom that in our enlightened society something like this could actually happen.

I wasn't the only one who believed that. In May, the UBS Securities Equity Research report announced, "A Win/Win Situation Appears Likely for CELLPRO; Rating Raised to Buy."

Even Wall Street was betting that we would survive.

12

WHEN I READ ABOUT PEOPLE DYING FROM lymphoma or leukemia or autoimmune diseases, I rarely make the comparison between their fate and my own. I think of my battle with lymphoma as a voyage that has been completed, but I live with the awareness of it and remain vigilant.

When I read about people dying from these diseases, I think of what might have been, and still wonder if there had been anything else I might have done to keep CellPro in business. In March 1999, a newspaper article reported the tragic story of a five-year-old girl suffering from aplastic anemia, a disease that destroys bone marrow, making it impossible for her body to produce new blood cells. "She needs a bone marrow transplant to survive past her sixth birthday," the reporter wrote, "but it's almost impossible to find a stranger to match a child who is a genetic melting pot of Cherokee, black, Hispanic and Jewish. . . . 'Eventually her system will rebel against the blood of strangers,' her doctor explained, further reducing the chances it will accept a bone marrow transplant. 'Do I think I'll find a match for her?' her doctor asks. 'No.' "

Sometimes I wonder what the people who put CellPro out of business would say if they read these stories. Whom would they blame? Would they still maintain it was the correct decision?

On July 24, 1997, Judge McKelvie handed down his final decision. I had anticipated the worst possible result, yet it was even worse than that. "Behind the science, medicine and the potential for treating patients," he wrote, "are investors who have demonstrated that their primary motivation is not humanitarianism, nor even responsible capitalism. The record in this case demonstrated that CellPro's motivation, as expressed by the words, conduct and testimony of its founders, is greed. They are prepared to stretch the boundaries of marketplace competition to maximize their

268

returns. They will deliberately take what is not theirs, pad their files and financial disclosures with weak and misleading opinions of counsel, and litigate to delay and frustrate."

In later writings, Judge McKelvie added that CellPro had "contempt . . . for the law; and for our system of civil justice"; that our lawyers presented "a weak pass at the quality of work one might expect from an independent counsel"; that the venture capitalists who funded Cell-Pro "hired counsel and bought opinions that the [Hopkins] patents were invalid"; and that those original opinions about the validity of Hopkins's patents were intended to "try to confuse or mislead . . . an unsophisticated jury."

When I conferred with Joe Lacob, he was irate. "When I read this opinion," he told me, "my blood just curdled. You can't sue a judge for libel, but if you could, I would."

But perhaps the angriest words came from Professor Martin Adelman, who is considered an expert on patent law and has testified as an expert witness in more than 150 patent-infringement cases—including our trial. "I think McKelvie's a good judge. I think this is the worst decision he's generated. . . . He criticized CellPro for being out to make money. . . . He just had a bad day. It's the stupidest thing I've ever read.

"I find his injunction . . . close to criminal. And I want to know how many people will die because of this, because this is not supposed to be what the system's about."

Harsh words, but I believed they were well-deserved. Judge McKelvie took it upon himself to reverse the unanimous decision of a jury, reinterpret Civin's patent claims, and then rule from the bench against us, based on his new interpretation. He denied us a new trial by declaring our potential defenses to these reinterpreted claims to be without merit, and inappropriate to be retried before a jury.

Judge McKelvie awarded the plaintiffs $6,961,479 in damages—the maximum amount permitted under the law, in addition to attorneys' fees—officially issued the injunction that would take the Ceprate system off the market as soon as an alternate system received FDA approval; prohibited us from exporting the device outside the United States after a one-year stay of the injunction; and, until the Isolex system was approved, ordered us to pay 60 percent of our sales to the plaintiffs.

It was a big victory for the interests of big business. Responding to this decision, Dr. William R. Brody, president of Johns Hopkins, said,

"Public and private research institutions are breathing a sigh of relief that such deliberate infringement of the patent system will not be tolerated by the federal courts. Patients gain from this decision as well. It is the very integrity of the patent system that fosters the large financial investments needed to bring medical advances through the development stage to reach the bedside."

Coincidentally, this same day McKelvie issued his final judgment, the FDA held its panel meeting to consider approval of the Isolex system. The data Baxter presented to the panel seemed woefully inadequate. It indicated that patients in the control group, patients whose stem cells had not been purified by the Isolex, actually engrafted faster than those who used it.

Reporting on this meeting, the Dow Jones news wire commented, "A confusing, split vote . . . has left the future of Baxter Healthcare Corp's Isolex 300 in question." Only six of fifteen panel members voted that failure to impair engraftment should be considered evidence of efficacy; meaning the fact that it did no harm should be sufficient reason to approve it.

Obviously the FDA could not approve the Isolex based on this hearing. Officially they reached "no consensus" on efficacy, but found it safe to use. They requested additional data from Baxter, which we believed was basically a means of saving face.

We immediately appealed McKelvie's decision. He was attempting to give to Baxter everything it had failed to get in negotiations or the previous trial. During the failed negotiations, for example, Baxter had demanded worldwide distribution rights for the Ceprate system, which was clearly an attempt to extend the reach of Civin's patent far beyond its legal limits. That was the basis of the antitrust lawsuit we had filed against them. But Judge McKelvie's injunction also applied to our operations in Europe, which seemed far beyond the limits of an American patent.

We appealed on numerous legal grounds. It seemed obvious to me that Judge McKelvie had made his decision months earlier, ignoring completely the findings of a jury after a trial, then set out to punish us by putting us out of business. Our hope was that the court of appeals would understand that we had been denied our basic right to a trial on the facts of the case.

Given the distance of passing time, I suspect it seems incredible that so many decisions could go against us and I would still maintain that we were absolutely right. When I became CEO at CellPro, my business philosophy was pretty simple: know at least as much as your competition does about your industry—preferably more—develop a product that people need, and continue to develop it. That was the model we followed, and it was right. On the business front, we beat Goliath. Knocked him out cold. But then Goliath brought in his lawyers.

We challenged the system—and the system won. In many of our battles, our opponents warned that we were a danger to the system. If CellPro were allowed to get away with this, corporations would no longer be able to depend on the sanctity of patents; if CellPro were allowed to get away with this, universities would lose the billion-dollar benefits derived from Bayh-Dole; if CellPro were allowed to get away with this, companies could earn huge profits from bogus patent opinions; if CellPro were allowed to get away with this, no patent would ever be beyond the reach of a competitor. All of these claims were ridiculous. We were not the slightest danger to the system. What we were asking for was consistent with the march-in provision of Bayh-Dole. We should have received the small-business preference and we did not, and the public health was potentially at risk because we were going to be forced out of business. Bayh-Dole set a very high standard, and we met it. There would never be a better case to be made for government intervention than to keep available to the public a technology developed with government money.

On August 1, 1997, NIH Director Dr. Harold Varmus denied our petition to march in. In his decision, Varmus claimed that he had not been persuaded that cell-separation machines had demonstrated any clinical benefit. "It is premature," he wrote, "for either Baxter or CellPro to claim patient benefits (other than a decrease in infusional toxicities)."

As I sat at my desk at CellPro reading these words, I thought about the woman who had been in the room next to mine at UW. "Other than a decrease in infusional toxicities"? I never learned what had happened to her after leaving the hospital, but I knew how she had suffered from infusional toxicities. I wondered what she might think about that statement. Even if that was the only area in which stem-cell separation proved beneficial, it should have been recognized as an important contribution. Perhaps, statistically, cell-separation technology had yet to demonstrate any clinical value, but I had been diagnosed with a terminal illness, and I was sitting at my desk at CellPro, bursting with life.

And deep inside I wondered if Dr. Varmus, or, for that matter, any of the other people who could have intervened along the way, might change their minds if they knew that in the future they would need a transplant and our technology would not be available.

Dr. Varmus did not believe that was possible. In his opinion he explained that even though stem-cell technology had not been proven to do any good, his decision ensured that patients would not be deprived of it. "The patient care implications of this matter were our first priority and concern. . . . Since both devices are currently available under the terms of the Court Order, I do not believe march-in proceedings are warranted. The NIH will continue to follow the situation to ensure that patient access to this technology is not compromised."

With this decision, the NIH pretty much surrendered its right ever to march in and disrupt the university-industrial partnership. If the government refused to march in in our case, it's difficult to imagine a scenario in which it would take action. Bayh-Dole had become too valuable to both universities and private industry for the government to interfere with it. A key protection for the public interest in Bayh-Dole had, for all practical purposes, been eliminated. There was no longer any guarantee that taxpayers would benefit from the research for which they were paying.

We were running out of survival options. It was tragic that after almost a decade our technology was blossoming, yet we had to begin drastic cutbacks. The value of our stock was falling rapidly. If I had paused even briefly to really consider the situation, I might have decided it was hopeless. Instead, as I had learned from Patty, I implemented.

The only real chance we had to continue as a viable company lay with the court of appeals. I knew we had a strong case. As one of the brokerage houses reported, "[A]s we reflected on Baxter's position, we grew more comfortable with our conclusion that CellPro's position on the appeal is persuasive." But by this point I had lost all faith that the institutions charged with protecting the public were capable of doing anything more than protecting themselves. Long ago, the concept of right and wrong had disappeared under the weight of procedural arguments. Long ago, concern for patients had ceased to play an important role in this drama.

Our challenge was to stay in business until the appeals court ruled. We immediately began cutting back on all our R&D. We continued those clinical trials already in progress, but, with only a few exceptions, curtailed all new projects. We allowed attrition to cut our staff from 160 to 120, and began making contingency plans for massive layoffs.

The first member of the Rick Project team to leave was Kirsten Stray. It was her choice. At the end of the summer she decided to join her husband in Utah. I wanted to see her before she left. Kirsten had been one of CellPro's first employees—she had been there when I joined the company—but we'd never spent any time together. Kirsten was an extremely quiet and unemotional person. When she did speak, she rarely said more than a few words. As I think back to the project team, I can recall the sound of everyone else's voice, but not Kirsten's. What I remember most is her smile, which sometimes seemed permanently affixed in place.

While there would always be a bond between us, that was left unspoken. She sat in my office and we spoke about her plans. Eventually I got around to thanking her for her participation in the project and told her, truthfully, how much she was going to be missed. "If you ever need a reference from me," I said, "you know you just have to ask." That seemed so little after all she had done. So I added, "And if for any reason you ever want to come back, you know you always have a job here." I began to realize there was no way I could adequately express my appreciation for her efforts, for all of their efforts. They were professionals. Just doing their job. I happened to be the very lucky recipient.

My son Ben was working at CellPro that summer, volunteering in the lab, when Kirsten happened to pass by. Ben and Kirsten had never discussed her role in the Rick Project. Ben wanted to bring up the subject, but he was hesitant. He wondered, How do you thank someone you really don't know for saving your father's life? Like father, like son.

Kirsten finally brought it up. "I'm glad everything worked out for your family," she said.

"I don't know how to really thank you for everything you did," he said. "But my family is very grateful." Finally, they hugged and she was gone.

As our successful European operations were not affected by McKelvie's decision, we considered spinning it off as an independent company. To make that a viable option, we rented lab space in Vancouver and in France. We immediately shipped our hybridoma and antibody to the lab in France.

We really needed both of these labs. We were still in the middle of numerous clinical trials, and we needed to continue to support Europe. McKelvie's injunction prevented us from doing the research necessary to test new antibodies and products in our own lab, so we set up a makeshift lab in a medical building across the border in Vancouver. It was only two hours from Seattle, but legally it was beyond McKelvie's reach.

Three or four CellPro employees would drive up there in a minivan on Monday and work in that lab all week, returning Friday to report their progress and discuss the next steps. I use the word "lab" quite loosely; it was actually two very small rooms with no air conditioning, modified to our specifications. We brought up a lot of equipment: columns, FACS equipment, centrifuges, a freezer in which to keep samples frozen. It felt very much like a clandestine operation.

When Baxter learned that we were producing our antibody offshore, Don Ware filed a motion requiring us to "repatriate," to bring back our antibodies to the United States from Europe, where the antibody would be destroyed "in the presence of a U.S. Marshal." These were vials of antibodies that were going to be used to treat patients, and they wanted them destroyed "in the presence of a U.S. Marshal." I was appalled when I read that motion, but certainly not surprised.

Baxter was scared that we were going to take our hybridoma offshore, out of the range of U.S. patents, and produce the antibodies needed to enable us to sell the system outside the country. We believed we had the legal right to do that, but McKelvie essentially granted their request.

The existence of that lab didn't really change anything, since realistically there was no chance CellPro could survive if we couldn't sell our product in the United States. Our European business was mainly marketing and distribution; we lacked the financial resources necessary to set up a manufacturing facility capable of supplying products and columns to the rest of the world. We would have had to start the company in Europe almost from scratch.

The noose was tightening. McKelvie's injunction required us to scale back by half our European sales, and to do it by February 1, 1998. That meant we would have to lay off our people there. We asked the appeals court for an expedited determination of the extraterritorial aspect of our appeal. We didn't believe an American patent could be extended beyond our borders into other countries. In this situation you have to demonstrate to the court that it is likely you are going to succeed on the merits of the case in this particular issue. After hearing our argument, the appeals court ruled that McKelvie's power did not extend beyond the range of U.S. patents, so they lifted that portion of the injunction.

A victory. We finally had a victory. The court of appeals does not like to reverse lower court judges, but the law was so clear that the court had no choice. While the immediate impact was important to us—it enabled us to stay in business in Europe—other people considered it a harbinger of the appeals court ruling on the entire case. Having reversed one part of the case, they reasoned, made it a lot easier to find fault with the rest of the decision. "It's a strong indication that the Court of Appeals is going to give it a good solid evaluation," decided the legal analyst for the brokerage house Auerbach Pollack & Richardson, who added that our brief "presents a powerful case for reversal of Judge McKelvie's decision."

We gained a second small victory two weeks later, when we received FDA approval to begin clinical trials of our B-cell depletion system. This approval came at a perfect time. Often we were so overwhelmed by our legal and business problems that we lost sight of the horizon. We forgot that the whole reason we were fighting these legal battles was to support the work being done in our labs. For the first time a tumor depletion system was going to be available to patients. It meant that other people were going to get the benefit of all the work done in the Rick Project.

The device had evolved slightly. It was a little slicker, a little easier to use, but essentially it was the same system that Nicole, Sharon, Kirsten, and Stan had created in eight incredible weeks.

In early April we enjoyed still another success as we met with the FDA Advisory Committee about expanding the legally permitted uses for the system. A decade's work was just beginning to mature. We had a workable platform, and we were finally ready to exploit it. The possibilities were many and exciting—all we had to do was stay in business.

Our stock rose literally and figuratively. The appeals court decision might have shaken the confidence of Baxter and Nexell, because they began hinting that we might be able to reach an agreement. Wouldn't it be to everyone's benefit if we could all come to the table and settle this thing, yadda, yadda, yadda? Maybe they meant it, or perhaps they had begun to realize that there were consequences to their actions. They had made claims to the court that they anticipated the Isolex device being approved by the FDA by the end of the 1997—but by the end of that year they still had not begun Phase III trials. It seemed apparent that while the Isolex eventually would be a useful cell-separation device, it lacked the flexibility of the Ceprate system. Baxter also had claimed that the two systems were essentially interchangeable and nothing would be lost if we went out of business. Maybe they were beginning to realize that just wasn't true.

We were prepared to put an offer on the table consisting of an up-front cash payment about equal to ten years' anticipated royalties—plus a royalty. It was far more than we had ever offered before, but it was the only means of guaranteeing we'd stay in business. We will never know what might have happened. Just as we were beginning settlement discussions, we were hit by a shareholder class-action lawsuit against us based on the court decision that we had willfully infringed the Civin patents.

Actually, the suit wasn't filed by our shareholders. With the growth of the biotech industry came the inevitable parasites: class-action lawsuits initiated by avaricious law firms. These law firms are corporate ambulance-chasers. They track a great number of stocks, and when there is a precipitous drop in one of them, they investigate the reason for it and, often on the flimsiest pretext, file a class-action lawsuit claiming mismanagement. These suits rarely go to trial. It costs management a lot less time, money, and energy to pay off these people than to wage a prolonged fight. There are few major companies in the biotech industry that haven't been involved in a class-action shareholder lawsuit. So it did not surprise me that soon after McKelvie announced his decision this lawsuit was filed against us.

The timing could not possibly have been worse. If we settled with Baxter before the appeals court ruled, the willful-infringement judgment would stand, making it very difficult for us to successfully fight the shareholders' lawsuit. We had no choice but to continue with our appeal.

No matter what steps we took to cut our burn rate, we were fast running out of money. We desperately needed to raise cash. Until this moment we had been able to avoid layoffs, but we no longer had a choice. In addition to reducing our cash burn rate by $300,000 to $500,000 monthly, downsizing the company significantly would send a signal to Wall Street that no matter what, we intended to see the patent case through to the end of the appeal. Combined with the positive signal sent by the appeals court when it stayed McKelvie's injunction, we thought we might be able to raise the money we needed to survive.

Downsizing. What an antiseptic term to describe the impact on a company and the lives of its employees when numerous people have to be let go. It's difficult for me to use the word *fired*, because that has come to imply that the employee has had some responsibility for losing his or her job. Nobody at CellPro was fired. Those jobs were lost because we couldn't afford to fund them anymore and remain in business, not because of any action of any person.

The layoffs started at the top. Larry Culver voluntarily decided to leave. If there is such a thing as business battle fatigue, Larry had it bad. He'd given everything he had to CellPro, and he just had nothing left. If a CEO is fortunate, he has one person on whom he can rely completely. I had Larry Culver. During my illness he had run the company without even encountering a speed bump. But it was time for him to leave, we both knew that.

One Sunday several months earlier, Patty and I had taken Larry and his wife, Shirley, sailing with us. I gave him the helm and set him up on the wind. We sailed north, into the wind. I remember looking at him as he had control of the boat. This was something he had never done before, and it was obvious he was loving it. He had a look of triumph on his face. But there was something else I noticed: he simply looked relaxed, he looked like he was having fun. All the tenseness had disappeared from his body. It had been so many years since I had seen him look that way that it surprised me. We'd been fighting so long and so hard that I'd forgotten completely what life once had been like. It pleased me greatly that I was able to give him that one day. But I knew then that Larry could not last much longer at CellPro.

We'd occasionally spoken about his leaving, and I'd always imagined it would happen at some unspecified time in the distant future. But when it became necessary to plan the layoffs, he told me, "You know, Rick, this really is the perfect time for me to go."

He was right, of course, but that made it no less painful. "You're probably right," I agreed. We'd become very close, so we often knew precisely what the other was thinking. I knew how difficult this fight had been on him. An extremely sensitive person, Larry had felt each blow struck at CellPro. It was very hard for me to lose him—he was my friend as well as the person on whom I relied—but I wanted to make this decision as easy as possible on him. "I understand," I said, "but, boy, I'm really gonna miss you. I don't know how I'm gonna hold up. Who am I going to talk to?"

We never formally said good-bye. We knew each other too well for that. I think if we had tried to have that parting scene, both of us would have cried, and neither of us wanted that.

Mark Handfelt, a young lawyer with shoulders broad enough to take the heat, replaced him.

I was exhausted, too, but even then I never considered leaving. I owed my life to CellPro, and I intended to be there until the day they pushed me out and locked the doors behind me.

A day or so after Larry left, I walked past his office. The door was open, and as I had been doing for so many years, I glanced in. The office was bare. The desk was cleared. I was momentarily stunned. The emptiness I felt wasn't just a reflection of that office. After everything Larry and I had been through together, all the battles we'd fought, all the late nights when we'd waded through the work, I felt very much alone.

At least Larry had made his own the decision to leave. And he had very quickly been hired by a telecommunications company. We were forced to cut forty people, most of them from research and development. We basically closed down all of our R&D. We had no choice. We had to continue manufacturing and technical support and sales, we had to continue all our clinical trials. About the only good news was that the biotech industry was flourishing in Seattle. There were a lot of good jobs available for good people.

Determining who would be laid off and who would escape these cuts was certainly one of the more difficult things I have ever done. We weren't working simply with lists of names; we were making determinations about the lives of friends, of people we knew. We knew their families. In many cases we had some idea of the impact this would make on their lives.

The actual decisions were made by the senior staff. We had many, many meetings at which we went over this numerous times until we finally had some idea who would be let go. Often it came down to choosing between two equally talented people. We offered the most generous severance packages possible under those conditions. But it was a hard time. We'd spent so long building this company, and now we were beginning to dismantle it.

I dreaded walking through the halls, fearing I might see someone who I knew was about to lose his or her job.

Among the jobs eliminated when we truncated R&D were those of Nicole Provost and Sharon Adams. But we didn't want to lose either of them. Joe Tarnowski asked both of them to come into his office. After they were seated, he said, "I need to talk to you."

Sharon looked at Nicole and said, "He's going to lay us off."

"No," Joe said, "but you're close." Nicole was being promoted to Director of Process Chemistry. Among her additional responsibilities would be running our quality-control program. Sharon would essentially be working with her.

Nicole was pregnant again, and she wasn't sure she wanted the job. "Look," she said, "why don't you just lay me off?"

Joe shook his head. "Sorry," he said, "and you can't quit, either. If you do, you don't qualify for the severance package." As the company contracted, multi-skilled people like Nicole and Sharon became invaluable. Both agreed to stay, at least temporarily.

As a manager, Nicole had to tell several members of her R&D group that they were being laid off. One of the names on that list was Stan Corpuz, the last of the original CellPro employees. Stan had helped build the prototype Ceprate system; he'd literally carried equipment from the basement laboratory in the old morgue into our first real offices and again into our own building. He'd probably done more jobs at the company than anyone else—including his membership in the Rick Project team.

Two years earlier, before his involvement on the team, before he found contentment with the Promise Keepers, this might have been very rough for him. But not now. Stan accepted the news with mixed feelings. While it was a shock, I think he also realized it was an opportunity. He had outgrown his job, and there was no place for him to go at CellPro.

"This isn't my decision," Nicole told him angrily, "and I won't defend it."

I was reading through some legal documents when Stan came to say good-bye. "Hey, come on in, buddy," I said with false bravado. This was a moment I'd known was coming, and feared. Exactly how do you say thank you very much and good-bye to a person who gave so much of himself to save your life? What are the right words?

Stan made it easy. "I just wanted you to know I'm proud of what we did here," he said. And then he thanked me for everything I had done.

He thanked *me*. He was worried about my feelings. It was an extraordinary display of grace at a most difficult time. I mumbled all the proper words, how much I appreciated everything he'd done for the company as well as for me personally, how we had no options, how this might even be better for him, call me if there's anything I can do to help find a new job. . . . And then we hugged. Those few seconds of intimacy said far more than all the compliments and reassurance I had tried to give him.

Stan told me he was actually looking forward to having the summer free to play golf. In fact, because our severance package was so good, some of the employees who lost their jobs referred to themselves as "winners of the CellPro lottery."

I had survived to preside over the death of CellPro. At times it was like watching a seemingly healthy person dying for no apparent reason, and being completely helpless to prevent it. Each morning when I went to work, I dug deep for fresh optimism. This was a preventable tragedy in the making, yet no one outside our company seemed to recognize it.

As a parting gesture, as they packed their belongings many people sent an e-mail message to the rest of the company. "I want to thank each of you for being generous over the past few years . . ." wrote Michael Jones. "I am grateful for the opportunity to have worked for this great company and with so many people I truly enjoyed. . . . I will miss all of you," sent Rita Lyons, one of several people who left voluntarily. "Well, the boxes are packed and I finally found that missing report," wrote Paul Harris. "Working here has been an enjoyable opportunity to work with excellent colleagues while doing something that helps people. . . ." And Jackie McGourty wished to all, "Stress not, sleep well, go well."

Nicole lasted about one week in her new job. Quality control is far more critical in biotech than in any other industry. Nobody's life is in jeopardy if a car's paint job isn't perfect. Quality control in biotech is a very laborious and time-consuming job, and carries with it tremendous responsibility. It's like climbing a paper mountain. After only a few days, Nicole realized it was not what she wanted to do. "Life's too short," she told Joe. "I just don't want to go through with this. I really think it's time for me to pack up."

Joe tried hard to talk her out of it. We need you, he told her, this is a very important job, we can—

"Joe, listen," she said. "In two months I'm having this kid. One way or the other you're going to have to deal with this at the end of the summer. And if I lose this kid I'm going to be really angry I stayed around here."

But leaving wasn't quite as easy as Nicole had imagined. She had made a substantial emotional investment in CellPro, and did not want her co-workers and friends to think she was bailing out in rough seas. We sat in my office one overcast afternoon, talking about just that. "That's not an issue," I told her truthfully. "I'm going to be forever grateful for everything you've done. But I want you to know, if we come through all this, after you've had some time at home on maternity leave, I'm going to come to your house and drag you back in here."

Eventually we compromised. We created a position for Nicole as a special consultant. She'd come in several days a week to write reports, attend meetings, assist with paperwork. But she had no specific responsibilities, nor did she supervise a staff or lead a group. On June 15, 1998, she sent Joe her official letter of resignation. ". . . I have truly enjoyed the challenges and rewards of working at CellPro for the past six and a half years, and I regret having to leave now when CellPro's fate still hangs in the balance. . . ."

Hanging in the balance? How about dangling over a precipice, holding on to a single unraveling thread? To raise cash, Mark Handfelt and I went back on the road to meet with investors. CellPro was a very high-risk investment, but there was a chance for an excellent payout. Our stock was selling for about four dollars. In terms of our profit potential, that was substantially undervalued. We were the only company with an approved product in a potential $100-million-a-year market. But the uncertainly over the lawsuit kept the price of our stock low. Investing in CellPro was a high-stakes gamble, but if our appeal was upheld, our stock would rise very quickly.

There were companies willing to bet on the outcome of the litigation. After reading the trial transcripts, an investment group led by Goldman Sachs was willing to invest $15 million in CellPro. But they wanted to wait to see how Wall Street reacted to the news that Larry Culver was leaving. When the market responded positively to our announcement that we were restructuring, Mark and I immediately flew back to New York to complete arrangements for the new financing. The deal was basically done. It was just a matter of doing the paperwork, filling in the blanks. Fifteen million was enough money to keep us in business through the appeal.

While we were meeting with Goldman Sachs in New York, Baxter's lawyers filed a motion in McKelvie's court that claimed we were squandering our resources, and asked the court to require us either to post a bond or prepay attorneys' fees in the amount of $8 million.

Once again, this was either an extraordinarily unfortunate coincidence—or Baxter had learned what we were doing and was trying to stop it. Baxter's lawyers claimed they were forced to make this motion when we announced our intention to reorganize, to ensure that we would have the money to pay the monetary damages McKelvie had levied against us. Perhaps that's true. Once again I have absolutely no evidence that it is not. But I probably can be forgiven for seeing Baxter's shadow behind every event.

Whatever their reasons, the motion served to scare away all our potential investors. No prudent banker would invest money that might have to be used to pay Baxter's legal fees. So over the next month we hammered out an agreement with Baxter in which we stipulated we would not conduct or undertake any transaction outside the normal course of business. Meaning we wouldn't try to hide funds, we wouldn't waste money, and we wouldn't simply spend it to keep their lawyers from getting it. In other words, we would continue doing business exactly as we had been doing it.

This agreement was acceptable to Baxter, Hopkins, and BD, as well as to the Goldman Sachs group. It looked as if we had successfully blunted another attack. Exactly one year earlier I'd been lying in a bed at the UW Medical Center being transplanted. I guess the good news was that since that time we'd been through so much at CellPro that my illness seemed like a distant memory, almost as if it had happened to someone else. But because of my illness my family had missed our annual summer sailing vacation in the Caribbean. These trips had become an important family tradition, and I didn't want to miss another one. I felt strongly that this vacation had been well earned by the Murdock family. So, with the Goldman Sachs deal seemingly in place we flew to the Caribbean. CellPro would survive to make the fight another day. My batteries needed recharging.

This was the maiden voyage of a magnificent forty-five-foot catamaran, which we'd christened *St. Somewhere,* from a Jimmy Buffett lyric. In addition to Patty and the boys, we had aboard four close Dutch friends. The eight of us spent many days sailing through the Virgin Islands.

The many battles of CellPro seemed a lifetime away from me. Once again I was a sailor, the captain of a wonderful new boat. My problems were limited to steering the right course and maintaining proper sail trim. I was a long way from the courtroom battles. There was nothing more I could do until the agreement we'd worked out with Baxter, BD, and Hopkins was translated into a court order. Then we would finalize the agreement with our new investors. By the time I got back, the deal should have been closed.

I was actually starting to feel optimistic. I'd brought a pager with me in case there was an emergency at CellPro, but I rarely phoned the office. I needed this vacation. On August 10, our lawyers filed papers outlining the new arrangement with the court in Delaware. All the concerned parties had agreed to the deal, and we were just fulfilling the legal formalities.

On August 11 we sailed into Cruz Bay in St. John. We went ashore for the day and separated to do some shopping or just wander around the harbor town. In the afternoon I found a pay phone and decided to call the office. When I reached my secretary, Connie Abella, she sounded relieved. "I'm glad you got my beep."

"I didn't, actually," I said, starting to get a little nervous. "I haven't been back to the boat for several hours. What's up?"

Mark needed to talk with me, she said. Not "wanted," needed. I was still pretty cheerful when he got on the line. "Hey, buddy, how you doing?"

"Not so good," he said somberly. "This is probably one of the worst days of my life."

We'd been through a lot of those days during the past few years. "What is it?"

"Rick, the appeals court decision came down this morning. We lost the case." We hadn't lost everything, just everything that mattered. The appeals court found that "the district court, when it construed the claims after trial, changed the rules of the game." But that did not seem to matter in their final judgment.

The court also reversed McKelvie on his attempt to extend the injunction beyond American borders, stating, "We agree with CellPro that the district court abused its discretion in ordering the repatriation and destruction of the exported vials." But they had upheld him on just about everything else. They agreed that the fact that Civin was never able to produce another antibody in his lab using the methods detailed in his patent claim could be excused because we failed to prove that the people working there were "of ordinary skill in the art" of making monoclonal antibodies. They agreed that McKelvie was not compelled to inform the second jury about the results of the first trial. And they agreed that Tom Kiley, one of the finest patent lawyers in the biotech industry, had erred by not recognizing shortcomings in Lyon & Lyon's original letters, "especially considering that the opinions did not express an opinion concerning infringement of the broadest claims."

With minor exceptions, the appeals court sustained McKelvie's decision. It was not an affirmation of his decision—that was not their function—but, rather, they upheld his legal right to make that decision. The appeals court looks only at errors of law. McKelvie had the legal right to reinterpret the patent claims, which he did—to fit Baxter's case. As long as he was acting within his powers, his rulings had to be upheld. Legally, we had willfully infringed the Hopkins patents.

I was long beyond any sense of disbelief at decisions made in a courtroom. Judges can find justification in the law for whatever decision they want to reach. This court of appeals happens to like Judge Roderick McKelvie. In fact, in several areas the court of appeals' decision was terribly inconsistent. For example, the bulk of the financial penalties levied by McKelvie were based on our profits outside the United States. While the appeals court held that McKelvie had no power beyond our borders, it did not rescind the penalties. Therefore we were being penalized for perfectly legal activities.

I could argue this case until my last breath, but it wouldn't matter. We were out of business.

It isn't often that life changes so drastically while one is standing at a pay phone on a Caribbean island. Although there were still several options available to us, I knew that the fight was done. Later that day I got a voice mail from Josh Green. "Well, Rick," he said with resignation, "this is it, baby. You fought the good fight. I'm sorry it turned out the way it did, but it's finally over."

I was shell-shocked. We hadn't expected the appeals court to rule for several months. And we certainly hadn't anticipated this ruling. At that moment there was nothing I could do. Flying back to Seattle would not have changed anything. I walked away from the phone in a daze. Ben was the first person I told, and he was devastated. He looked as if someone had hit him with a sledgehammer. He had spent the summer working at CellPro, he knew all the people, he'd worked in the "war room" during the first trial and had been living this battle with me, day after day. Ben was still young enough to have faith that the system would recognize right and wrong, so this really shook him up. He asked me many questions, but I had no answers for him.

I slung my arm over his shoulder and we walked through the small village. Eventually we found Patty. When I told her the news, at first she was incredulous. Then she became totally enraged. Rarely, if ever, had I seen her so furious. She cried with anger. How could this happen, she asked, how could the legal system deprive patients of potentially lifesaving technology? What was wrong with these people? I had no answers for Patty, either. Years later I still don't.

I never drink to smooth distress. I generally limit myself to one good gin and tonic. But not that night. That night I had no limits. There were just so many thoughts rumbling wildly in my head; within minutes my mood would change from utter depression to a sense of relief. But there was one thing I felt very deeply: I had done everything possible to save CellPro. I would be shoved out the door with my head held high.

Coincidentally, Nicole happened to be at CellPro the day the appeals court decision was announced. As she remembers it, it was as if the entire company breathed a sigh of relief. Finally we knew our fate. We had been treading water much too long, unable to reach the safety of the shore, yet unwilling to give up. Everyone was very sad about the outcome and anxious about his or her own future, but no one cried. At least now there would be some rest.

There were many details still to be worked out, including putting together a good severance package for our employees, but essentially CellPro was dead. We could have gone further in the appeals process. In fact, one brokerage house reported, "We think that the decision is illogical in several respects and therefore the judgment has an above-average chance of further review," but we were spent emotionally as well as financially. The fight was out of us.

Prior to leaving on this vacation, I had gone in for a full post-transplant workup. My platelet count, which had remained low, was slowly edging up toward normal. My CT scan was clean. All that was left was my bone-marrow biopsy. As each test was completed, Sherri Bush would call me with the results: "Everything's clean, Rick." "Looks real good, Rick." Ollie and Sherri were most concerned about the CT scan, which was perfect, but I worried about the bone marrow. The results of the bone-marrow tests were not ready when we'd left, so Ollie told me he would leave me a voice mail as soon as he got them. But as the days passed, there was no message from him.

There was no reason to believe there was any abnormality in my marrow, but the fear that my lymphoma might recur lives in a dark little place in the back of my mind. And there were moments when I began to wonder if the news had been bad and Ollie simply did not want to ruin this family vacation.

That lurking shadow allowed me to keep the disastrous news about CellPro in perspective.

The day I got back to my office, I found an e-mail from Ollie. It had been sent two days after I'd left on vacation. Ollie does much of his communication by e-mail. He'd simply forgotten to leave me a voice-mail message. "I just want to pass on your bone marrow results to you," it read. "They are completely clear." Not that I ever doubted that, of course. Of course.

There were so many things that had to be done, and done fast, when I got back. But even as I threw myself headlong into those tasks, I knew I was falling onto a thick, soft mattress; whatever else happened, I was healthy.

We had one last gasp left. One of the two companies that had purchased a license for the Civin patents from Baxter was SyStemix, which is now a subsidiary of the giant Swiss pharmaceutical company Novartis. SyStemix had never developed a cell-separation product, but they still owned that license. They didn't have the right to sell it to us, but if they bought us, we could apply their license to the Ceprate system and continue in business. For a brief time it looked as if we might be able to make a deal. They were intensely interested. CellPro made a nice fit with their existing oncology business. I flew down to the San Francisco Bay area to meet with their management. "This is a real business," I told them, "and it's growing very fast. This is a way for you to acquire approved technology for a fraction of a penny of what it's actually worth."

We asked for $15 million, but would have taken substantially less. Whatever we received would have meant a small payout to our stockholders, but, more important, CellPro would have remained in existence—

perhaps under another name—our people would have kept their jobs and the technology would live on.

Several Novartis executives came to Seattle from Basel, Switzerland. I wanted to believe this could happen, that this story would have a happy ending. But Novartis was also cutting staff, and several of the executives who lost their jobs were the people with whom we were meeting. SyStemix's CEO really agonized over this, but there just wasn't enough time to put together a deal. Just a few days more and it might have happened. But we didn't have a few days.

Ironically, the only company able to save the Ceprate system technology was Baxter. In late August, representatives of Baxter and Nexell met with Mark Handfelt, Joe Tarnowski, and several other corporate officers to try to negotiate a sale of the entire business, everything from our technology to our employees. Except me, of course, and the senior management team. Baxter and Nexell indicated in writing they might be willing to purchase the company for as much as $10 million, but it soon became apparent they had no plan at all. After six years of litigation they didn't have the slightest idea what they wanted to do with CellPro. Their only interest, as it turned out, was to keep the Ceprate system on the market until their Isolex system was approved by the FDA. Eventually, Handfelt met with their representatives in New York—and they offered $1 million for our intellectual property. That was all they wanted. They had no interest in our ongoing business. Eventually we sold the remnants of CellPro to the plaintiffs for about $3 million worth of VimRx stock.

Shutting down CellPro made no business sense at all. I remember Joe Tarnowski just shaking his head in amazement when he heard that news. As Joe said, "What are they thinking? They're letting our best asset, our experience and intelligence, walk out the door. They can take all the paper they want, but there's going to be no one here to translate it for them."

They could easily have just walked in and taken over the entire company. That certainly would have been the best thing for patient care. We estimated we were about a year away from being profitable, so they certainly might have very rapidly recouped whatever they paid. There were some people there who understood that.

In conversations with Tarnowski, some Nexell executives privately questioned the wisdom of closing CellPro. We had an FDA-approved product being produced in an FDA-approved facility. We had other extremely promising products in clinical trials. We had a proven platform technology easily adaptable to numerous applications. And we had an extraordinary staff and a first-rate clinical development organization. Nexell was still far away from its first approved product. And even when the Isolex was approved, the Ceprate system was much easier to use in the clinic. Simply, we were light-years ahead of them.

So why would they simply trash everything we'd accomplished? Why would they just throw away this valuable technology? I'm sure they can justify their actions. But to me it was simply the end result of a vendetta. Baxter had long wanted CellPro—and me—to disappear. They considered me the devil incarnate because I'd stood in front of TV cameras during our march-in campaign and bloodied them badly. I'd told the world what I considered to be the truth, and they were very bitter about that. They made it very clear to members of our board that they wanted nothing to do with me. Whatever happened to CellPro, Murdock had to go. Now that they had the opportunity, they were going to do it.

The community of transplanters was perhaps more surprised than we were at the decision of the appeals court. We'd been living through this possibility for several years, but as doctors they really didn't believe such a promising technology would actually be taken out of their hands. It seemed too absurd to be believable that a business dispute could affect the lives of so many seriously ill patients. This wasn't the promise of some valuable treatment to be delivered at some vague point in the future, but an existing technology that was being taken away from them. They were using it every day, lives were being changed, and—poof!—it was gone. As our medical director, Amy Sing, once said, "As a scientist and a physician, to have it come down to a dispute over the antibody being used is astounding. How can you patent biological systems? The fact is that we had a very different core technology."

Almost immediately after the decision of the appeals court was announced, our customers began ordering every disposable kit we had available. Each kit is one patient treatment. Richard Burt ordered eighty kits, for example. Within days we sold our entire inventory. Anxious doctors were calling to tell us they had patients scheduled months in advance and needed additional kits. We tried to accommodate people wherever we could, but sometimes it was impossible. We made a deal with Nexell to provide eight hundred disposable kits that theoretically would have allowed them to treat patients until the Isolex received approval. We also gave all our clinicians a thirty-day notification to end all the trials in progress. We provided the materials they needed to complete the treatment of all their scheduled patients, but told them they could not initiate any new treatments—they couldn't treat anyone else.

One of the first things Nexell did, according to Amy Sing, was raise the price of the Isolex system from $3,750 to $5,000, and the price of the Ceprate system disposable kit, which they were distributing, from $4,300 to $5,000. She also anticipated that the price of the Isolex would be raised after it received FDA approval, as is normal procedure. Well, thanks to the court and the NIH, they now have a monopoly in this business, so they can pretty much charge any price they desire. And free from a competitive marketplace, they can choose to move forward in any direction they choose.

I went to work each day as if I were in a daze. But for the most part I slept well at night. What I felt wasn't exactly a sense of relief, but rather acceptance that the battle was finished. And there were so many things to be done. It was important to me that we be able to offer our employees a decent severance package, and after some legal maneuvering we were able to do just that.

I received my first telephone call from a headhunter only hours after the appeals court decision was announced. The first decision I made as I began my search for a new position was that it would have to be in a medical area in which Baxter did not have a presence. The anger and bitterness were too deep. Almost immediately I received several outstanding offers. One of the most flattering came from Dr. Richard Burt at Northwestern. "You know," he told me, "this gene therapy work we've been doing is starting to move ahead pretty fast. I'd like you to consider coming here and starting a company with me."

For a time I didn't want to do anything. I wanted the luxury of days to think.

Unlike a store, we couldn't simply lock the door and walk away. There was still considerable corporate housekeeping that needed to be done before CellPro ceased to exist, but most of that was taken care of by Mark Handfelt. Over several weeks the halls became silent, the memos and Dilbert cartoons disappeared from the cubicles, the now-useless computers were piled up. And in the storeroom, brand-new Ceprate systems that would never be used sat neatly on the shelves. I remember standing outside the storeroom door one day, looking through a glass window at a pile of Ceprate devices that would never be shipped, and thinking, *About all they're good for now is boat anchors.*

The sense of finality was overwhelming. Walking through the corridors was like patrolling the deck of a ghost ship. And there remained one thought shared by everyone: *what a waste, what an incredible waste.*

When I look back at everything that happened—and I do, occasionally—I think about those opportunities I might have had to make a deal with Baxter. I'm as guilty as anyone for believing that eventually we would reach a fair accord with Baxter, that experienced executives on both sides would remember that we were a medical device company serving a population of daring transplanters and seriously ill patients. We weren't making widgets, we were in the business of saving lives—mine, for example—and somehow that should have made a difference.

It's impossible ever to really know how much was lost forever the day we closed the doors. At the very least, I agree with the declaration of Dr. Fred LeMaistre of the South Texas Cancer Institute, who wrote, "I believe as a practical matter, withdrawal of the CellPro device . . . could harm and even lead to death in a significant number of patients."

I know the cost in lives that might have been saved is a great one. As Coe Bloomberg told me once, referring to the second trial, it's going to get a lot worse before it gets better. About a year after the final verdict, on August 6, 1999, I was sitting at my desk at my new company reading a front page story in the *Wall Street Journal* titled "How a Corporate Feud Doomed Human Trials of Promising Therapy: Baxter Beat CellPro in Court, Some Say Dying Patients Are the Case's Big Losers."

"After four failed bone-marrow transplants," the article began, "Stephen Grupp's patient, an eight-year-old girl with advanced leukemia, died in January.

"Dr. Grupp looks frustrated as he describes the harrowing struggle. 'In the end,' says the pediatric oncologist at Children's Hospital (in Philadelphia), 'we just ran out of options.'

"There is one option Dr. Grupp did not get to try: an experimental procedure that is aimed at some of the deadliest of all cancers. It's so new that researchers are still trying to figure out if it works. Yet in patient trials last year 15 children with advanced leukemia—all of whom were regarded as terminal—a third using the treatment survived bone-marrow transplants with no sign of cancer afterward. A larger set of trials was to begin last November on 50 children with advanced leukemia and two groups of adults. . . .

"The trials never took place."

While the rest of the article outlines the dispute without making any determination of the blame, it makes absolutely clear the fact that this was a tragedy that should have been avoided.

That child's death was only the beginning. Doctors are admitting it now, people who might have lived will die. What was lost forever in that courtroom was not only our approved system, but the medical advances we might have made. We were on a trail and, no, we'll never know where it might have led.

The Ceprate system is gone, and with it the proven avidin/biotin cell-separation technology. It would require tens of millions of dollars for another company to even begin to develop it again. The competition that would have spurred further development of transplant technology, gone. The working relationships we established with our clinical investigators, gone. The B-cell depletion device that saved my life and was being used in clinical trials, gone. The T-cell depletion device being used in pivotal Phase III clinical trials to mediate GvHD, gone. The breast cancer program, gone. Multiple myeloma, childhood leukemia, and gene therapy programs, gone. The use in other autoimmune diseases as multiple sclerosis, rheumatoid arthritis, and lupus and AIDS, gone. The potential for new therapeutic approaches using stem cells, gone. Our cancer vaccine program being conducted at six sites and about to start in twelve others, gone. The whole new series of applications just over the horizon for cell therapy, gone.

Even assuming that the Isolex works as well as the Ceprate system, years of research and patient care are gone.

In May 1999 the Isolex system finally received FDA approval, perhaps more rapidly than it might have if CellPro had remained in existence. Neither Baxter nor Nexell has ever completed a Phase III randomized controlled trial with the Isolex system. They received their approval based on Phase I and II data, meaning they were clearly held to a different standard than CellPro in the approval process. The question, of course, is why? As a condition of our final settlement, Baxter made us withdraw our citizen's petition to the FDA protesting this special treatment; I'm convinced that if we hadn't done that, the Isolex system still would not be approved.

At their annual stockholders' meeting in May 1999, VimRx announced that it was changing its name to Nexell and had acquired Baxter's minority interest, although apparently Baxter would retain responsibility for manufacturing, sales, and distribution of the cancer-related cell-therapy products. So, after this long fight, Baxter is no longer even in the stem-cell business. How sad.

It's difficult to learn what products Nexell intends to produce. At that same meeting the president and COO, L. William McIntosh, said, "Although the Isolex systems will play a key role in cell selection for stem cell transplantation following high dose chemotherapy, these applications are just the tip of the iceberg. Already our cell selection and expansion technologies are being used in novel treatment approaches to autoimmune diseases, genetic abnormalities, viral infections and other serious life-threatening diseases. We view our technologies as the critical path to realizing the potential of using living cells to cure disease, and the road ahead is very exciting indeed."

Exactly as I would have said, but I would have said it many years earlier.

On its Web site, Nexell boasts, "The newest [Isolex] combines both positive and negative in one automated procedure. With this procedure median T-cell depletion of 5 logs can be achieved." Additionally, they are apparently in Phase III clinical trials for autologous transplants for breast-cancer patients, they are supporting allogeneic transplants for diseases beyond cancer, evaluating stem-cell transplants for the treatment of autoimmune diseases, and participating in an NIH gene-therapy study.

Nexell also announced, "With the diagnostic products acquired from CellPro, Nexell is launching a new diagnostic division to support the development and marketing of innovative diagnostic technologies. . . . We plan to launch our first diagnostic product in 1999."

Compared with the more than sixty clinical trials in which we were actively involved, Nexell claimed to be supporting only twenty clinical studies. As regards Nexell taking over our diagnostics program, there is no increased benefit to the patient in Nexell's trying to run a program developed by our scientists using our technology.

A lot of the programs in which we were actively involved seemed to have disappeared. To be fair to Nexell, which rightly claims to have been totally uninvolved in the patent-infringement litigation, I have tried to find out about other programs not mentioned in their publicity releases, but I have not been successful. "With regards to your interest in our clinical programs," Nexell president and CEO L. William McIntosh responded to my request, "I must refer you to information that is in the public record. As you can imagine, details of our product and clinical development program are considered confidential and, should we disclose any details, it must be disclosed publicly in fairness to investors and in compliance with SEC laws and regulations. However, be absolutely assured that we have done everything humanly possible to insure that investigators and clinicians worldwide have access to cell selection technology in order that important work, including that formerly sponsored by CellPro, can and is continuing to move forward."

So there may well be programs in progress of which I am not aware. But I do know, as reported by the *Wall Street Journal*, that patients who might benefit from the treatment I received cannot get it, that other extremely important trials have been halted, and that children who might have lived have died. I do know that.

And I do know, again as reported by the *Journal*, that the NIH, which refused to "march in" because, against our warnings, they accepted the promise that patient care would not be compromised, supposedly is beginning to investigate the cause of this situation. I must wonder, though, how thorough that investigation can be if neither I nor anyone at CellPro has been contacted and knows nothing about it.

What's been lost? Programs, people, knowledge, lives, innovation, the incentive of competition, an incredibly promising future, and time. At the very least, time. Even if someday Nexell is able to duplicate what we accomplished, which I believe is impossible, the world of transplantation has been set back years. Years. And as someone who faced a terminal disease cured by a transplant knows so well, to a patient nothing is more valuable than time.

What's been gained? For 1998, Nexell reported revenues from product sales of $13.4 million on $19.2 million in end-user sales—which seems to mean that Baxter earned approximately $6.2 million, or almost one-third of Nexell's revenues, as its manufacturing and distribution margin.

In his decision, Judge McKelvie wrote, "CellPro argues that the terms of the permanent injunction will harm its ability to survive as a company and to sell products that would serve the public health interests . . . [but] CellPro's shut-down arguments are speculative at best."

Perhaps now Judge McKelvie will accept that we were serious.

CellPro exists now as a corporate shell—investors bought it for the tax benefits. I accepted the job of CEO and president of Kyphon, an innovative orthopedic medical device company developing a new technology to be used to treat osteoporotic spinal fractures. Patty and I moved to the San Francisco Bay Area and live on the water. Jamie has graduated with his masters degree in biomedical engineering. Ben is completing his senior year at Bates College. His Senior thesis is about the use of immunotoxins to treat cancer. I find the fact that both of my boys will end up working in the medical field very satisfying. Patty has become enchanted with the Bay Area, and is painting and gardening and exploring. We sail often now.

Only weeks after the appeals court reached its final decision, Nicole Provost gave birth to a daughter, Alexa, a six-pound, one-ounce bundle of love.

And what about Baxter and Becton Dickinson and Johns Hopkins? When it became publicly known that I was going to tell this story, the president of Crown Publishers received a letter—with substantial enclosures—cosigned by counsel for Baxter, BD, and Johns Hopkins that argues their side of the story. As they emphasize, they won in a courtroom and that verdict was affirmed in the court of appeals. To the credit of the publisher, I was not even told about this letter until informed of its existence in correspondence with Nexell. The letter, in its entirety, is included in the appendix of this book. While their lawyers are correct in stating the result of the litigation, I lived this story quite differently. I believe that this letter tells the true story of what happened about as accurately as their statement that my life was saved by "stem cell purification technology invented at Johns Hopkins." In fact, the only time the facts in this case were presented to a jury, that jury found in our favor on all counts.

A more accurate assessment of the case, I believe, was contained in a letter written to the *Seattle Post-Intelligencer* by a CellPro stockholder named Eric Cohen. In response to a series of articles on the demise of CellPro, Mr. Cohen wrote:

> *The fact is that CellPro won the patent suit. At least by what is generally considered an inviolable right of our society, a trial by jury. Judge McKelvie threw that jury verdict out and replaced it with his own personal verdict, but only after broadly rewriting the patents, at the*

behest of Johns Hopkins, Becton Dickinson and Baxter, to justify his ruling. Then, after raging at CellPro, their attorneys and management, he slapped an illegal injunction on their international sales that was devastating (even the appellate judges ruled the court "abused its discretion"). The entire four-year history here is a stunning aberration from anything considered "normal" in a courtroom and the P-I missed the mark by not examining further the actions and motives of the federal judge.

For me, the actions of Judge McKelvie will forever remain a mystery. I will never forget the incredible adventure I lived through at Cell-Pro, although I don't often have the time to reflect on it. I'm extremely busy at Kyphon, putting to work the hard lessons I've learned, but sometimes a reporter will call with a question about CellPro, or I'll read a story about an advance in stem-cell technology. (Supposedly, for example, Canadian researchers grew an entire mouse from a single stem cell.) When that happens I'll lean back at my desk and clasp my hands behind my head and just wonder what might have been.

The fact that I am alive and capable of that wonder is the miracle of it all.

And if I ever doubt that the fight for CellPro was worth it, I have in my possession an e-mail reading, "I just want to say thanks. I am one of the 131 test patients with multiple myeloma. I was the first to be treated at the hospital of the University of Pennsylvania and now, approaching three years, there have been no myeloma cells detected. I am working sixty hours a week, ride bikes and horses, playing tennis and doing whatever I damn well feel like doing. I feel this would not have been possible at this time without your machine. I will be forever thankful."

This is the way science is done in America.

APPENDIX

THE LETTER

We are writing on behalf of The Johns Hopkins University, Becton Dickinson and Company, and Baxter Healthcare. We understand that Crown Publishers has purchased the right to publish a book called "The Experiment" to be co-authored by Richard Murdock, the former president of CellPro, Inc. According to the publicity about the book, it will cover not only Mr. Murdock's battle against mantle-cell lymphoma, but also "his fight against a mammoth, multinational medical products corporation that was using archaic patent laws to threaten the survival of his company." We assume that the latter statement refers to the patent infringement litigation brought by our respective institutions ("plaintiffs") against CellPro in the United States District Court for the District of Delaware.

At the outset, let us say that we have no objection to the publication of a book about Mr. Murdock's experience as a patient undergoing stem cell transplantation in connection with his treatment for a rare form of cancer. Mr. Murdock's successful recovery has vividly illustrated the tremendous potential of the stem cell purification technology invented at Johns Hopkins, and the widespread publicity his treatment received has helped to bring this breakthrough technology to the attention of clinicians and patients.

To the extent the book also discusses patent matters, however, we want to be sure that Crown Publishers has access to all the facts and that it takes seriously its obligation to present the facts accurately and to avoid false and defamatory statements about our respective institutions. For this reason we are attaching copies of several pertinent court decisions, which contain extensive findings of fact based on documentary evidence and sworn testimony.

The first opinion is the district court's June 28, 1996, decision. This decision describes the background of Johns Hopkins inventions in the field of stem cell transplantation; these inventions, the court found, represented "a major breakthrough in medical science." In recognition of this, the U.S. Patent and Trademark Office awarded Johns Hopkins four patents, two of which were asserted against CellPro. In its decision, the district court determined, as a matter of law, that the use of CellPro's Ceprate SC stem cell purification device infringed one of Johns Hopkins' patents (the '680 patent) and accordingly overturned a jury's verdict to the contrary. The court's decision also granted plaintiffs a new trial on all other issues that previously had been tried to the jury, on the ground that the jury's verdicts in favor of CellPro were "against the great weight of the evidence" presented at trial.

The next opinion, dated July 24, 1997, considers and rejects CellPro's contention that enforcement of the patents against CellPro constituted "patent misuse." In this opinion, the court made findings of fact that on three separate occasions in 1992, Baxter offered CellPro an unconditional patent license under the Johns Hopkins patents. The terms offered to CellPro were the same terms offered to and accepted by other companies in the field; only CellPro turned them down. As the court's findings demonstrate, the three offers to CellPro did *not* condition Baxter's grant of a patent license to CellPro on CellPro's grant of product distribution rights to Baxter, contrary to CellPro's repeated assertions.

The third opinion, also dated July 24, 1997, addresses the assertion that CellPro proceeded with development and marketing of its Ceprate SC device because it "believed" the Johns Hopkins patents were invalid and that CellPro's device did not infringe them. This specific question was put to a jury in March 1997, which found unanimously that this assertion was untrue. In the district court's written opinion on the same issue, the court made detailed findings of the fact that CellPro knowingly, willfully, and with no good faith basis infringed on the Johns Hopkins patents. The court of appeals also affirmed the district court's decision to award plaintiffs treble damages, an amount the trial judge found was appropriate "to punish CellPro for its deliberate and bad-faith infringement of the [Johns Hopkins] patents."

We trust that Crown Publishers will review the district court's findings of fact carefully, recognizing that such findings, affirmed on appeal, constitute definitive judicial determinations of the facts, and that those determinations are legally binding on CellPro and its officers and directors (including Mr. Murdock) in any subsequent proceedings. As the opinions demonstrate, they were based on the court's detailed and thorough consideration of the evidence, which included the internal notes, memoranda, and the trial testimony of numerous CellPro employees, including Mr. Murdock, as well as CellPro's confidential FDA filings, its SEC filings, and other testimonial and documentary evidence.

If we can be of any further assistance, including providing copies of relevant court decisions or copies of evidentiary materials that underlie the findings and conclusions of the trial and appellate courts, please contact our outside litigation counsel, Donald R. Ware. . . .

INDEX

ABOUT THE AUTHORS

RICK MURDOCK earned a degree in zoology from University of California Berkeley. In 1991 he joined CellPro as vice-president of marketing and business development. He is now president and CEO of Kyphon, Inc., a medical-device company in the San Francisco Bay Area, where he lives with his wife, Patricia.

DAVID FISHER is the author of more than 40 books. His best-sellers include *Gracie* and *All My Best Friends* with George Burns, *The Umpire Strikes Back* with Ron Luciano, and *Conversations with My Cat*. He is the only author to have a novel, a work of nonfiction, and a reference book, *What's What,* offered simultaneously by the Book-of-the-Month Club. He lives in New York and Atlanta with his wife, Laura, two boys, two cats, and a very noisy dog.

Please visit Rick and David at their Web site, www.PatientNumberOne.com, with your comments, or to view additional materials.